From Counting to Continuum

Understanding the natural numbers, which we use to count things, comes naturally. Meanwhile, the real numbers, which include a wide range of numbers from whole numbers to fractions and exotic numbers like π, are, frankly, really difficult to describe rigorously. Instead of waiting to take a theorem–proof graduate course to appreciate the real numbers, readers new to university-level mathematics can explore the core ideas behind the construction of the real numbers in this friendly introduction. Beginning with the intuitive notion of counting, the book progresses step-by-step to the real numbers. Each sort of number is defined in terms of a simpler kind by developing an equivalence relation on a previous idea. We find the finite sets' equivalence classes are the natural numbers. Integers are equivalence classes of pairs of natural numbers. Modular numbers are equivalence classes of integers. And so forth. Exercises and their solutions are included.

EDWARD SCHEINERMAN is Professor of Applied Mathematics and Statistics at Johns Hopkins University. He is the author of various books including textbooks, a research monograph, and a volume for general readership, *The Mathematics Lover's Companion*. He has twice been awarded the Mathematical Association of America's Lester R. Ford Award for outstanding mathematical exposition and has received numerous teaching awards at Johns Hopkins. His research publications are in discrete mathematics.

"This book takes us on a fascinating journey through the world of turning intuition into rigor. Deep and elegant ideas are presented at just the right level of detail to keep the reader interested and engaged. A perfect introduction for anyone who is open to seeing the beauty of mathematics."

– Maria Chudnovsky, Princeton University

"Providing a careful, rigorous construction of the field of real numbers is among the greatest intellectual achievements in human history. This outstanding book will take you on an engaging and exciting guided tour of the real numbers that explains the mysteries and conveys the magic of this beautiful conceptual foundation for mathematical analysis."

– Johns Hopkins University Applied Physics Laboratory

"Scheinerman is a master expositor who here presents a patient and thorough entry into the world of real numbers, providing sufficient precision and detail to make the definitions mathematically correct but in a way that is accessible to a wide audience."

– David Bressoud, Macalester College

From Counting to Continuum
What Are Real Numbers, Really?

EDWARD SCHEINERMAN
Johns Hopkins University

Shaftesbury Road, Cambridge CB2 8EA, United Kingdom

One Liberty Plaza, 20th Floor, New York, NY 10006, USA

477 Williamstown Road, Port Melbourne, VIC 3207, Australia

314–321, 3rd Floor, Plot 3, Splendor Forum, Jasola District Centre, New Delhi – 110025, India

103 Penang Road, #05–06/07, Visioncrest Commercial, Singapore 238467

Cambridge University Press is part of Cambridge University Press & Assessment, a department of the University of Cambridge.

We share the University's mission to contribute to society through the pursuit of education, learning and research at the highest international levels of excellence.

www.cambridge.org
Information on this title: www.cambridge.org/9781009538640

DOI: 10.1017/9781009538657

© Edward Scheinerman 2025

This publication is in copyright. Subject to statutory exception and to the provisions of relevant collective licensing agreements, no reproduction of any part may take place without the written permission of Cambridge University Press & Assessment.

When citing this work, please include a reference to the DOI 10.1017/9781009538657

First published 2025

A catalogue record for this publication is available from the British Library.

A Cataloging-in-Publication data record for this book is available from the Library of Congress.

ISBN 978-1-009-53864-0 Hardback
ISBN 978-1-009-53867-1 Paperback

Cambridge University Press & Assessment has no responsibility for the persistence or accuracy of URLs for external or third-party internet websites referred to in this publication and does not guarantee that any content on such websites is, or will remain, accurate or appropriate.

To Jonah

Contents

Preface   xi

0   Prelude   1
    Exercises 5

1   Fundamentals   7
    1.1   Sets 8
    1.2   Lists 10
    1.3   Relations 10
    1.4   Equivalence Relations and Partitions 13
    Exercises 16

2 $\mathbb{N}$: Natural Numbers   23
    2.1   You Can Skip this Chapter, but Don't 23
    2.2   Starting from Nothing 24
    2.3   Stuff to Count 25
    2.4   Prelude to Counting: Sets of the Same Size 26
    2.5   Finite Sets 28
    2.6   Natural Numbers are Equivalence Classes 29
    2.7   Addition 31
    2.8   Multiplication 35
    2.9   Less-than-or-equal 37
    2.10   The Peano Axioms 40
    2.11   Primes 42
    Exercises 46

3 $\mathbb{Z}$: Integers   53
    3.1   Reintroducing the Integers 53
    3.2   An Equivalence Relation for Pairs of Natural Numbers 54
    3.3   Integers and their Operations 56
    3.4   Arithmetic Properties of the Integers 61
    3.5   Assimilation 63
    Exercises 63

4	**$\mathbb{Z}_m$: Modular Arithmetic**	**67**
	4.1 A Relation Based on Divisibility	67
	4.2 Modular Arithmetic	69
	Exercises	71
5	**$\mathbb{Q}$: Rational Numbers**	**73**
	5.1 An Equivalence Relation for $\mathbb{Z} \times \mathbb{Z}^*$	73
	5.2 Rational Arithmetic	75
	5.3 Order Properties of the Rational Numbers	77
	5.4 Assimilation	78
	Exercises	79
6	**$\mathbb{R}$: Real Numbers I, Dedekind Cuts**	**81**
	6.1 The Case for Real Numbers	81
	6.2 Left Rays and Real Numbers	83
	6.3 Addition	86
	6.4 Less-than-or-equal	90
	6.5 Multiplication	91
	6.6 Completeness	97
	6.7 Assimilation	99
	Exercises	99
	Addendum: Density of Rational Squares	102
7	**$\mathbb{R}$: Real Numbers II, Cauchy Sequences**	**105**
	7.1 Cauchy Sequences	105
	7.2 Equivalent Cauchy Sequences	109
	7.3 Addition and Multiplication	111
	7.4 Ordering Real Numbers	113
	7.5 Bisection and our Favorite Real Number	115
	7.6 Assimilation and Completeness	118
	Exercises	120
8	**$\mathbb{R}$: Real Numbers III, Complete Ordered Fields**	**123**
	8.1 Isomorphism	123
	8.2 All Complete Ordered Fields are Isomorphic	127
	8.3 The Real Numbers have a Square Root of 2	128
	Exercises	130
	Addendum: Cauchy Sequences and Completeness Revisited	132
9	**$\mathbb{C}$: Complex Numbers**	**135**
	9.1 From i to $\mathbb{C}$	135
	9.2 Complex Numbers as Equivalence Classes	137
	9.3 Polar Coordinates	139
	9.4 The Fundamental Theorem of Algebra	143

| | Exercises | 149 |
| | Addendum: Complex Exponential | 150 |

10 Further Extensions   **153**
    10.1 Infinities 153
    10.2 Quaternions 159
    10.3 p-adic Numbers 160
    Exercises 167

Answers to Exercises   **171**
    Chapter 0 171
    Chapter 1 172
    Chapter 2 176
    Chapter 3 180
    Chapter 4 184
    Chapter 5 185
    Chapter 6 188
    Chapter 7 192
    Chapter 8 196
    Chapter 9 200
    Chapter 10 203

Bibliography   **207**

Index   **209**

Preface

Let's begin with a provocative claim: *Real numbers are useless.*

No real-world problems require real numbers for their solution. Whether determining how much mortgage a home buyer can afford, prescribing a medication dosage for a patient, or measuring the trajectory of a space probe, the rational numbers do perfectly well. One may require just a few digits of accuracy or perhaps a dozen digits to the right of the decimal point. Not enough? It is not likely there's a real problem to be solved that requires precision to 50 or more digits. Real numbers are irrelevant to the real world. Even if we reject this heretical assertion, it's possible to use real numbers without even knowing what they really are.

Do people *need* to know what a real number really is? Perhaps not, but I expect some *want* to know.

This is an interesting endeavor. Real numbers are not easy to define, and it's an important accomplishment of generations of mathematicians that we can give a precise definition.

The focus of this book is on *definitions* and much less on theorems and their proofs. An A-to-Z completely rigorous development of the real number system is for upper-level mathematics majors in an analysis course. However, the core ideas can be enjoyed by a broader audience, and those are the readers I hope to serve.

Audience

The objective of this book is to present a basic introduction to defining the real numbers, $\mathbb{R}$. The reader will come away understanding that: there are complete ordered fields and all complete ordered fields are isomorphic; the real numbers are any one of these; and everything we need to know about the real numbers can be derived from the fact that $\mathbb{R}$ is a complete ordered field.

This book is aimed at self-learners, armchair mathematicians, mathematics students who want to understand what numbers actually are, and people who want a user-friendly path into mathematics. It would be a perfect book to read before, during, or after taking a course in calculus or real analysis, or for use in a seminar course.

From *Saturday Morning Breakfast Cereal* [12].

No calculus is required for this book, but readers need to be comfortable with pre-calculus-level mathematics.

It is worth noting that this book is not a book on the foundations of mathematics. We don't start with the Zermelo–Fraenkel axioms and we make no mention of the Axiom of Choice. We don't prove everything we assert. We don't distinguish between *sets* and *proper classes* when we partition all finite sets into equivalence classes. We don't begin by setting $0 = \emptyset$, $1 = \{\emptyset\}$,

$2 = \{\emptyset, \{\emptyset\}\}$, and so forth.

To do everything "right" would mean not reaching readers who can still profit from the ideas in this book.

To the Reader

This is a book about numbers. We begin with the most familiar numbers: 0, 1, 2, and so forth. These are exactly the numbers you need to answer questions of the form "How many?" We use these basic building blocks to carefully define the integers, then the rational numbers, then the real numbers, and – as a bonus – the complex numbers.

You have already learned something about the real numbers through decimal notation. That is, a real number consists of a (typically infinite) stream of symbols chosen from this collection:

$$- \quad . \quad 0 \quad 1 \quad 2 \quad 3 \quad 4 \quad 5 \quad 6 \quad 7 \quad 8 \quad 9$$

Certain sequences are valid such as −3.501 or 22, but others are nonsensical such as 3.3..2 or 53−.

Some real numbers have exactly one representation in this system, such as

$$\pi = 3.14159265358979323846264\ldots.$$

but others have more than one representation:

$$1 = 1.00000\ldots = 0.9999\ldots.$$

You've likely been using this notation long enough that it feels comfortable. Let me try to make you feel unsettled. When we add or multiply decimal numbers, we begin at the right. To multiply 37×18 we first calculate $7 \times 8 = 56$. We write down the 6 and carry the 5. Here's the first step:

$$\begin{array}{r} 37 \\ \times \quad 18 \\ \hline 6 \end{array}$$

Likewise, when we add $37 + 18$ the first step is to calculate $7 + 8$; we always begin at the right.

So how do we calculate (say)

$$\pi \times e = 3.14159265358979323846264\ldots \times 2.718281828459045235360\ldots$$

when we can't go all the way to the right?

> ### You Know What a Number Is. Can You Explain It?
>
> Augustine of Hippo wrote:
>
> > *Quid est ergo tempus? Si nemo ex me quaerat, scio; si quaerenti explicare velim, nescio.*
> >
> > What then is time? If no one asks me, I know. If I want to explain it to someone who asks, I do not know.
>
> If no one asks you what a *number* is, it feels like you know. What happens, then, if you are asked to explain what a numbers is to someone else? Does an explanation evade you?

One of the questions we address in this book is: What is the square root of 2?

Here's a worthless answer: The square root of 2 is a number that when multiplied by itself gives the result 2.

Why is that a bad answer? Suppose you asked me: What is the secret to a happy life? Imagine I replied: It's hidden knowledge that makes your life happy. Are you happy with that answer?

A better answer to the $\sqrt{2}$ question is this: The square root of 2 is 7/5. This is an incorrect answer, but it's a pretty good wrong answer because

$$\frac{7}{5} \times \frac{7}{5} = \frac{49}{25},$$

which is nearly 2.

A terrific answer to "What is the square root of 2?" would be a fraction a/b where a and b are whole numbers (integers). We could check if the answer is correct by calculating

$$\frac{a}{b} \times \frac{a}{b} = \frac{a^2}{b^2} = 2.$$

Alas, as you may be aware, there is no such fraction.

To then say "Ah yes, but $\sqrt{2}$ is an irrational number" is just a fancy way to say $\sqrt{2}$ is not a rational number (is not a fraction). That says what it isn't. What is it? And how do we know there is any such number? I could begin to write down an infinite decimal number (complete with three dots) and say *that* is the square root of 2, but how would we know it's correct? Can you really multiply this number

1.4142135623730950488016887242096980785696716875376948073...

by itself and see that you get exactly 2? Besides the fact that I haven't shown you most of the number, how could you ever do a calculation on an infinite string of digits?

There is a square root of 9: $\sqrt{9} = 3$ because $3 \times 3 = 9$. Is it possible there is no $\sqrt{2}$?

The good news is that there is a square root of 2. The difficult, but really interesting, fact is that $\sqrt{2}$ is a *real number*, but saying precisely what a real number is isn't easy. It's a long journey from 1, 2, 3 to $\sqrt{2}$, but it's one I hope you enjoy.

The Content of this Book

We start with the notion of *counting*. We present the natural numbers as the answers to counting questions. In Chapter 2 we present the concept of *finite set* and define two finite sets to be equivalent if there is a one-to-one correspondence between them. Natural numbers are then defined as equivalence classes of finite sets.

This is a paradigm that is repeated throughout the book. We begin with a basic structure (e.g., finite sets). We present an equivalence relation on those structures. We then create a new family of numbers as equivalence classes of those structures. We reinforce this with a typographical convention. In the chapter where they are defined, the newly created numbers (equivalence classes of simpler objects) are presented in **boldface**. See the boxed comment on page 28.

Thus in Chapter 3 we extend the natural numbers by creating an equivalence relation on pairs of natural numbers; the equivalence classes are the *integers*.

We take a brief detour in Chapter 4 to create modular integers as equivalence classes of integers.

In Chapter 5 we extend the integers by creating a new equivalence relation on pairs of integers; those equivalence classes are *rational numbers*.

Chapters 6 and 7 give two different definitions of the real numbers. In Chapter 6 we define real numbers as equivalence classes of left rays of rational numbers (Dedekind cuts). Chapter 7 constructs real numbers as equivalence classes of Cauchy sequences of rational numbers.

The tension of having two different definitions of real numbers is resolved in Chapter 8, in which we give the most important definition in this book: the real numbers are a complete ordered field. We explain the notion of isomorphism and assert that all complete ordered fields are isomorphic.

Each step in this journey is motivated by a "failure" of a given set of numbers. Natural numbers fail to provide subtraction. Integers fail to provide division. Rational numbers fail to provide a square root of 2. It is reasonable to note that the real numbers also have a "failure": There is no square root of -1.

Chapter 9 rectifies this last failure by extending the real numbers to the complex numbers. Not only do we show that all complex numbers have square roots within the complex numbers, but we present the Fundamental Theorem of

> **Numbers as Equivalence Classes**
>
> | Numbers (chapter) | Raw materials | Equivalence relation |
> |---|---|---|
> | $\mathbb{N}$ (2) | Finite sets | Bijection |
> | $\mathbb{Z}$ (3) | $\mathbb{N} \times \mathbb{N}$ | $(a,b) \equiv (c,d)$ iff $a+d = b+c$ |
> | $\mathbb{Z}_m$ (4) | $\mathbb{Z}$ | $a \equiv b \pmod{m}$ |
> | $\mathbb{Q}$ (5) | $\mathbb{Z} \times (\mathbb{Z} - \{0\})$ | $(a,b) \equiv (c,d)$ iff $ad = bc$ |
> | $\mathbb{R}$ (6) | Left rays | $|L \triangle L'| \leq 1$ |
> | $\mathbb{R}$ (7) | $\mathbb{Q}$-Cauchy sequences | See §7.2 |
> | $\mathbb{C}$ (9) | $\mathbb{R}[x]$ | $p(x) \equiv q(x) \pmod{x^2 + 1}$ |
>
> A recurring motif in our development of $\mathbb{R}$ is to create each new type of number as equivalence classes of more basic objects. This table summarizes our approach.

Algebra and its proof (omitting some topological rigor) to make the case that $\mathbb{C}$ is a reasonable end to the journey.

That said, as a bonus we have Chapter 10 in which we give a gentle introduction to a variety of more exotic concepts of number such as the extended reals, tropical arithmetic, hyperreals, quaternions, and p-adic numbers.

Acknowledgments

I received wonderful input and feedback from colleagues and students that helped me tremendously. I especially want to thank Tanuj Alapati, Tontong Chen, Gabriel Gormezano, Stephen Kennedy, Evan MacMillan, Daniel Naiman, Daniel Scheinerman, and Jennifer Song.

Many excellent suggestions came from anonymous reviewers, and I appreciate their input.

Thank you to Mairi Sutherland for her enormously helpful copy editing, and to Kaitlin Leach, Arman Chowdhury, Laura Emsden, and their colleagues at Cambridge University Press – it has been a pleasure working with you!

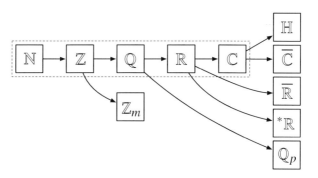

This diagram shows the dependence of the many type of numbers presented in this book. Our primary focus is the progression $\mathbb{N} \to \mathbb{Z} \to \mathbb{Q} \to \mathbb{R} \to \mathbb{C}$ but we also touch on a few other types of numbers on our journey.

Chapter 0

Prelude

Before we set off on our journey from counting numbers to real numbers, we need to say a bit about mathematics as a discipline.

Mathematical work is a three-legged stool that rests on these key notions:

- **Definitions**: In mathematics we require utterly unambiguous descriptions of concepts.

- **Theorems**: These are statements of mathematical facts that we know, with absolute certainty, to be true.

- **Proofs**: These are unassailable arguments used to demonstrate that theorems are true.

Metaphorically speaking, *Theorem* and *Proof* are the Oscar-winning movie stars of mathematics. It's a great honor to have one's own name attached to a theorem, such as the Pythagorean Theorem in geometry.

Definitions are often relegated to a less glamorous supporting actor role. In this book, however, our focus is on definitions, especially on the definition of the real numbers. Nevertheless, we need to also be mindful of theorems and their proofs. Let's say a bit more about these three concepts.

Definition

A fabulous feature of mathematics is the ability to be utterly precise. In some realms it is difficult to give iron-clad definitions, especially when personal judgment comes in to play. It's no joke that it's difficult to define precisely what it means for something to be *funny*.

For example, what does it mean for a number to be *even*? Sometimes it's enough to give several examples and hope that the idea is clear. We could say, "The even numbers are 2, 4, 6, 8, and so forth." Is that clear? What about 0? What about −2? Are they even?

We can do better. A definition gives a precise criterion for the concept being introduced. Consider this:

> **Definition 0.1.** An integer n is *even* provided $n = 2a$ for some integer a.

From this we see that 0 is even because $0 = 2 \times 0$. Likewise, -2 is even because $-2 = 2 \times (-1)$. On the other hand, 3 is not even because there is no integer a so that $3 = 2a$; the only number a that satisfies that equation is $a = \frac{3}{2}$, which is not an integer.

Theorem

A *theorem* is a mathematical assertion that is proven to be true. Mathematicians use a variety of other terms for theorems including *proposition*, *lemma*, and *corollary*. Typically the word *theorem* lends an air of importance and depth, where as a *proposition* is more modest.

For example, the following is merely a proposition:

> **Proposition 0.2.** *Let a and b be even integers. Then a + b is also even.*

On the other hand, the following is more substantive and therefore is called a theorem:

> **Theorem 0.3** (Pythagoras). *Let a and b be the lengths of the legs of a right triangle and let c be the length of the hypotenuse. Then $c^2 = a^2 + b^2$.*

In both cases, the statements are absolutely true without any exceptions. We know they are true because we can prove them, and that leads us to a discussion of proof.

Proof

In many disciplines the truth of an assertion is established by looking at many examples and ascertaining that a certain pattern holds reliably. From experience, I can promise you that the following is true: *During July, the weather in Baltimore is hot and humid.* No one would dispute this, even though once in a great while Baltimoreans enjoy a cool, dry July day.

Mathematics is much less nuanced. When we assert, as we do in Proposition 0.2, that the sum of two even integers is even, we mean that it is so in every possible case. How can we know? There are infinitely many possibilities! The answer is we are able to write a *proof* of this claim.

A *proof* is an absolutely irrefutable argument that a mathematical assertion is true. Often this is accomplished by logical deduction from the definitions. Here's an example.

Proof of Proposition 0.2. Let a and b be even integers. By Definition 0.1 we know that $a = 2x$ and $b = 2y$ for some integers x and y.

Note that $a + b = 2x + 2y = 2(x + y)$. Therefore $a + b$ is $2z$, where z is the integer $x + y$. Thus, by Definition 0.1 we see that $a + b$ is even. □

A more interesting type of proof is *proof by contradiction*. The key idea in such a proof is that we *suppose* that the conclusion to the proposition is false and then argue that such a supposition leads to an impossible conclusion. It then follows that the supposition must be wrong, and that the conclusion to the proposition is true.

Let's look at two examples to get a sense of how this works. Let's begin with a simple statement:

Proposition 0.4. *Let a and b be nonzero real numbers. Then ab is not zero.*

To prove this, we have to prove that ab is not zero. In a sense, we'll ask: Is it possible that $ab = 0$? What happens if that's the case? We show that it leads to a contradiction. Here's how this works:

Proof of Proposition 0.4. Let a and b be nonzero real numbers. Suppose, for the sake of contradiction, that $ab = 0$.

Since $a \neq 0$ we know that a has a multiplicative inverse, a^{-1}. Therefore

$$a^{-1} \cdot (ab) = (a^{-1} \cdot a)b = 1 \cdot b = b.$$

We are given that $ab = 0$ and therefore $a^{-1} \cdot (ab) = a^{-1} \cdot 0 = 0$. From this it follows that $b = 0$.

But we are given that $b \neq 0$ which contradicts the conclusion that $b = 0$. This is impossible. Because the supposition that $ab = 0$ leads to a contradiction, we conclude that $ab \neq 0$. □

Perhaps Proposition 0.4 is too basic to be impactful. Let's look at another example that may be less familiar.

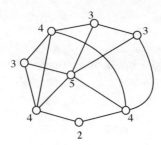

Figure 0.1: A simple graph. The number next to each vertex is its *degree*: the number of edges incident with that vertex. Note that some vertices have the same degree.

A *graph* is a mathematical structure[1] consisting of nodes (called *vertices*) and connections between pairs of nodes (called *edges*). A graph is called *simple* if the edges always join distinct vertices (no edge joins a vertex to itself) and between a given pair of vertices either there is no edge or there is exactly one edge (there cannot be two or more edges joining the same pair of vertices). An illustration of a simple graph appears in Figure 0.1.

The *degree* of a vertex in a graph is the number of edges attached to that vertex. Every vertex in Figure 0.1 is labeled with its degree.

In a simple graph with n vertices, the degree of a vertex can be any number from 0 (if the vertex is not joined to any others) up to and including $n - 1$ (if the vertex is joined to all the others).

Notice that in Figure 0.1 that there are some vertex degrees that are repeated: There is more than one vertex with degree 3 and more than one with degree 4. However, there's only one vertex each with degree 2 and 5.

Must there always be a repeat? The answer is yes.

Proposition 0.5. *In a simple graph with two or more vertices, there must be at least two different vertices with the same degree.*

Let's see how to prove this using proof by contradiction. We need to prove that there must be a repeated degree. If that's not so, it means that all the vertices have different degrees from each other. Here's the proof.

---
[1]This is not an iron-clad mathematical definition. Our objective is to illustrate the concept of proof by contradiction, not to delve into the details of graph theory.

Proof of Proposition 0.5. We are given a simple graph with n vertices where $n \geq 2$. Suppose, for the sake of contradiction, that all the degrees of the vertices are different from each other.

The degrees of the n vertices are values from 0 up to $n-1$. Since that's exactly n different numbers (and there are no repeats), it must be the case that there is one vertex with each of these degrees: 0, 1, 2, and so on, up to $n-1$.

This means there is a vertex a that has degree 0. It is not joined by an edge to any other vertex.

This also means there is a vertex b that has degree $n-1$. It must be joined to all the other vertices.

Are a and b joined by an edge?

- No, because a is not adjacent to any others.

- Yes, because b is adjacent to all the others.

This is a contradiction. Therefore our supposition that all the vertex degrees are distinct is impossible. It follows that there must be a pair of vertices with the same degree. □

Exercises

0.1 What does it mean for an integer to be a perfect square? Write down a precise definition.

Begin like this: An integer n is called a *perfect square* provided . . . .

0.2 The Fibonacci numbers are

$$0, 1, 1, 2, 3, 5, 8, 13, 21, 34, 55, \ldots.$$

Write down a definition for the Fibonacci numbers that begins like this: The *Fibonacci sequence* is the list of numbers in which . . . .

0.3 What does it mean for a triangle to be equilateral? Write down a precise definition.

0.4 Let A, B, and C be three points in a plane. What does it mean for them to be collinear? Write down a precise definition.

0.5 Define what it means for two lines to be parallel without referring to distance. Give two definitions: one in which you assume the two lines are in a plane and one in which the lines are in three-dimensional space.

0.6 Prove that the product of two even integers is even.

0.7 An integer n is called *odd* provided there is an integer k such that $n = 2k+1$.

Prove that the sum of two odd integers is even and that the product of two odd integers is odd.

0.8 Proposition 0.4 asserts that if a and b are nonzero real numbers, then ab is nonzero. Show that this conclusion does not work if a and b are nonzero *matrices*. That is, find nonzero matrices a and b for which ab is the zero matrix.

Chapter 1

Fundamentals

Mathematics starts with numbers, and this is a book about numbers, from the simple to the complex (pun intended). We begin this journey with the *natural numbers*. These are the familiar numbers 0, 1, 2, and so on, and the set of all natural numbers is denoted by $\mathbb{N}$.

Natural numbers are the answers to counting questions. You know how to count. Given a jar full of marbles, you take out a marble and say aloud "one." Then you take out another and say "two." You continue in this fashion until the jar is empty and your final utterance is the number of marbles. See Figure 1.1.

Undoubtedly you understand these simple numbers well. Nevertheless, we take a careful look at what it means to count and, in so doing, introduce core ideas from mathematics.

Counting, as illustrated in Figure 1.1, involves creating a *one-to-one correspondence* between *sets*. In the figure, one of the sets is the marble collection and the other set contains the numbers 1, 2, 3, 4, 5, and 6. We examine these ideas and other concepts as background before we begin our journey to the real number system.

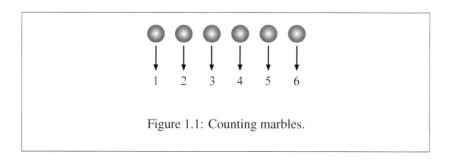

Figure 1.1: Counting marbles.

> This chapter introduces *equivalence relations* and *equivalence classes*. These concepts are important in every chapter of this book, and so we urge you to be sure your understanding of these ideas is rock solid before proceeding to subsequent chapters.

1.1 Sets

A *set* is an unordered collection of things without repetition. A set may be presented by listing its members between curly braces, like this: $\{1, 2, 3, 4, 5, 6\}$.

When we say that a set is an *unordered* collection we mean that the sequence in which the elements are listed between the braces is irrelevant. The set $\{1, 2, 3, 4, 5, 6\}$ is exactly the same as the set $\{2, 4, 1, 6, 5, 3\}$.

When we say that a set is a collection *without repetition* we mean that an object either is or is not a member of the set; the object cannot be in the set more than once.

The members of a set are called *elements* of the set. The symbol $\in$ is used to indicate when an object is a member of a set. Specifically, if x is an object and A is a set, the notation $x \in A$ means that x is an element of the set A. For example, if A is the set $\{1, 2, 3, 4, 5, 6\}$ then $2 \in A$ is true. However, $7 \in A$ is false and we write this as $7 \notin A$.

Empty and Singleton Sets

The simplest set of all is the *empty set*; this is the set that has no members. The notation for the empty set is $\varnothing$. The empty set is also called the *null set*.

A *singleton* is a set with exactly one member, for example $\{9\}$ is a singleton set.

While the definition of a *singleton* is simple enough, let's take another approach in describing a singleton without using numbers. The reason for this alternative approach is a desire (in Chapter 2) to use singletons to give meaning to the number 1. In other words, we present a definition of *singleton* that does not rely on any concept of number.

> **Definition 1.1.** A set A is a *singleton* provided:
> - A is not empty,
> - if $x \in A$ and $y \in A$, then $x = y$.

The first condition implies that there is at least one element in A, and the second condition implies that there can't be two (or more) different elements.

We see that these two descriptions of *singleton* amount to the same thing, but the second avoids any mention of numbers.

Set Operations

There are two core set operations: *union* and *intersection*.

The *union* of sets A and B is a set that contains all objects that are a member of A or a member of B (or both). The union of these sets is denoted $A \cup B$.

For example, suppose the two sets are

$$A = \{1, 2, 3, 4, 5, 6\} \quad \text{and} \quad B = \{2, 4, 7, 9\},$$

then $A \cup B = \{1, 2, 3, 4, 5, 6, 7, 9\}$.

The *intersection* of sets A and B is the set of elements that are in both A and B. The notation for the intersection of sets is $A \cap B$.

For the same sets A and B above, we have $A \cap B = \{2, 4\}$.

We say that a pair of sets are *disjoint* provided they have no elements in common. That is, A and B are disjoint sets provided $A \cap B = \emptyset$.

Finally, if A and B are sets, the *difference* $A - B$ is the set of all elements of A that are *not* in B. Reprising the example above, suppose

$$A = \{1, 2, 3, 4, 5, 6\} \quad \text{and} \quad B = \{2, 4, 7, 9\},$$

then $A - B$ is the set $\{1, 3, 5, 6\}$. Elements 2 and 4 are in A but because they are also in B they are excluded from the difference. Note that $B - A = \{7, 9\}$.

When B is a subset (see below for the definition) of a "master" set U, the difference $U - B$ is referred to as the *complement* of B.

Subsets

Given sets A and B, we say that A is a *subset* of B provided every element of A is also an element of B. The notation is $A \subseteq B$.

For example, if A is the set $\{1, 3\}$ and $B = \{1, 2, 3, 4, 5, 6\}$ then $A \subseteq B$, that is, A is a subset of B.

However, with $A = \{1, 3, 9\}$ and $B = \{1, 2, 3, 4, 5, 6\}$ it is not the case that A is a subset of B because 9 is a member of A, but not of B.

It is important to understand the difference between $\in$ and $\subseteq$. When we write $a \in A$ it means that a is a member of the set; it does not mean that a is a subset of A. These examples should help clarify:

- $2 \in \{1, 2, 3, 4, 5\}$ is true: the number 2 is an element of the set $\{1, 2, 3, 4, 5\}$.

- $\{2\} \subseteq \{1, 2, 3, 4, 5\}$ is also true: the set $\{2\}$ is a subset of $\{1, 2, 3, 4, 5\}$ because all elements of $\{2\}$ (there's only one!) are also members of $\{1, 2, 3, 4, 5\}$.

- $2 \subseteq \{1,2,3,4,5\}$ is false because 2 is an element, not a subset.
- $\{2\} \in \{1,2,3,4,5\}$ is also false.
- Let A be the set $\{1, 2, \{3\}\}$. Note that, in this case, 3 is not an element of A and $\{3\}$ is not a subset of A. The elements of a set may, themselves, be sets. In this case $\{3\} \in A$.

For any set A, it is always true that $A \subseteq A$ because all of A's elements are, of course, in A. A *proper* subset of A is a subset of A that is different from A.

1.2 Lists

A *list* is an ordered collection of things in which repetition is permitted. A list is presented by writing its elements between parentheses, like this: $(1, 3, 2, 2)$.

Lists are *ordered*. That means that $(1, 3, 2, 2)$ and $(1, 2, 3, 2)$ are different. For two lists to be considered the same, the members of the two lists must be identical and in the same sequence.

Repetition is permitted in lists. This means that $(1, 3, 2, 2)$ and $(1, 3, 2)$ are both valid lists, but they are different.

In this book we are especially interested in lists with exactly two elements; these are called *ordered pairs*. Some examples: $(1, 2)$, $(2, 1)$, and $(3, 3)$.

Ordered pairs are not sets: $(1, 2) \neq \{1, 2\}$.

Two ordered pairs are equal provided their first entries match and their second entries match. That is, $(a, b) = (c, d)$ exactly when we have both $a = c$ and $b = d$.

1.3 Relations

Mathematics uses a wide variety of relations between objects. A *relation* (also called a *binary relation*) is a predicate – a kind of sentence – that compares two things. For example, $\subseteq$ is a relation between sets; for some pairs of sets it is true and for other pairs it may be false:

- $\{1, 2\} \subseteq \{1, 2, 3, 4\}$ is true,
- $\{1, 2, 3, 4\} \subseteq \{1, 2\}$ is false.

In addition to is-a-subset-of, we also encountered $\in$, the is-an-element-of relation. This relation compares objects with sets.

There are many other familiar relations; here are some examples:

- $=$, equality: When we write $x = y$ we are asserting that x and y are exactly the same.
- $\leq$, less-than-or-equal, used to compare numbers.

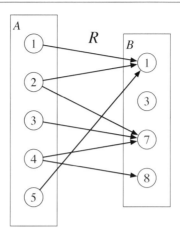

Figure 1.2: An example of a relation R from the set $A = \{1, 2, 3, 4, 5\}$ to the set $B = \{1, 3, 7, 8\}$. An arrow $a \to b$ in this diagram indicates that $a \; R \; b$.

- $\cong$, is-congruent-to, used to compare geometric figures such as triangles.

We can define our own relations. Specifically, given two sets A and B, we define a relation R from A to B by specifying which pairs $a \in A$ and $b \in B$ are such that $a \; R \; b$ is true. As an alternative to listing all the pairs (a, b) for which $a \; R \; b$ is true, we can draw a picture as in Figure 1.2.

In the diagram the elements of the two sets are represented by circles, and arrows indicate the relation; that is, there is an arrow from an element $a \in A$ to an element $b \in B$ to indicate that $a \; R \; b$. For the relation visualized in the figure, we see that $1 \; R \; 1$, $2 \; R \; 1$, and so forth.

Notice that $3 \; R \; 7$ is true for this relation, but $7 \; R \; 3$ is false. Indeed, there is no value a in A such that $a \; R \; 3$; this is apparent in the figure because there are no arrows pointing to 3 on the right (in set B).

One-to-one Correspondences

Can we count without numbers?

Given a set A, we might ask: How many elements are in A? For example, A might be the collection of marbles in a jar. Counting the elements of A is tantamount to finding a *one-to-one correspondence* between the elements of A and the numbers 1, 2, 3, and so forth. This is illustrated in Figure 1.1 on page 7. Alas, this action requires that we already have numbers at our disposal.

Still, we can describe when two sets have the same number of elements without using the word *number*. Instead, we rely on the idea of a *one-to-one correspondence*.

For example, consider these sets:

- {apple, banana, cherry},
- {Argentina, Botswana, Canada}, and
- {anger, bliss, confusion, dread}.

We can demonstrate that the first two sets have the same number of elements without using numbers (or saying the word *three*) by pairing fruits with nations like this:

apple $\to$ Argentina,   banana $\to$ Botswana,   cherry $\to$ Canada.

However, it is easy to see that there is no way to pair the emotions set with the others.

Definition 1.2. Let R be a relation from a set A to a set B. We say that R is a *one-to-one correspondence from A to B* provided all of the following are true:

- For every element $a \in A$, there is a $b \in B$ such that $a \, R \, b$ [existence].
- For every element $a \in A$, if $a \, R \, b$ and $a \, R \, b'$ then $b = b'$ [uniqueness].
- For every element $b \in B$, there is an $a \in A$ such that $a \, R \, b$ [existence].
- For every element $b \in B$, if $a \, R \, b$ and $a' \, R \, b$ then $a = a'$ [uniqueness].

Thinking about relation diagrams, the first pair of conditions says that for each $a \in A$ there is exactly one arrow emanating from a. The second pair of conditions says that for each $b \in B$ there is exactly one arrow pointing to b. Notice, however, that Definition 1.2 does not rely on any notion of number.

Said yet another way, a one-to-one correspondence from A to B gives a matching between each element of A with a unique element in B. The implication of this is that the two sets, A and B, have exactly the same number of elements.

With this language in place, we see that counting the elements of A means finding a one-to-one correspondence between A and a set of the form $\{1, 2, \ldots, n\}$.

1.4 Equivalence Relations and Partitions

In this section we focus on a particular type of relation known as an *equivalence relations*.

Definition 1.3. Let R be a relation from a set A to itself. We say R is an *equivalence relation* on A provided it satisfies the following three properties:

- **Reflexive property**: If $a \in A$ then $a \, R \, a$.
- **Symmetric property**: If $a, b \in A$ and $a \, R \, b$, then $b \, R \, a$.
- **Transitive property**: If $a, b, c \in A$ with $a \, R \, b$ and $b \, R \, c$, then $a \, R \, c$.

For any set A, the is-equal-to relation ($=$) is an equivalence relation. Indeed, the word *equivalence* is meant to evoke the image of equality. However, equality is not the only equivalence relation.

Let A be the set of triangles in the plane and let $\sim$ be the is-similar-to[1] relation. For example, any two equilateral triangles are similar to each other, and hence satisfy this relation. It is easy to see that any triangle is similar to itself [reflexive], if triangle T_1 is similar to triangle T_2, then we also have $T_2 \sim T_1$ [symmetric], and if $T_1 \sim T_2$ and $T_2 \sim T_3$, then $T_1 \sim T_3$ [transitive].

Here's another example from geometry. Let A be the set of lines in the plane and let $\|$ be the is-parallel-to relation. This, too, is an equivalence relation.

Exercises 1.20–22 ask you to show that *has-a-one-to-one-correspondence-with* is also an equivalence relation. This fact is important for Chapter 2.

Equivalence Classes

Let R be an equivalence relation on a set A. An important observation is that R slices A up into pieces; the elements in each piece are all related to each other by R, but not to any others. Let's look at this precisely.

Definition 1.4. Let R be a relation on a set A and let $a \in A$. The *equivalence class* of a is the set of all elements of A that are related, by R, to a. In notation:
$$[\![a]\!]_R = \{x \in A : x \, R \, a\}.$$

---

[1] Reminder: Two triangles are *similar* if the angles in one are the same as the angles in the other. Thus, any two triangles with, say, angles 20°, 50°, and 110° are similar.

The notation $[\![a]\!]_R$ stands for the equivalence class of a. When the relation R is unambiguous from context, the subscript may be omitted.

The expression $\{x \in A : x \, R \, a\}$ is an instance of *set builder notation*; see Exercise 1.15. In words, it says: This is the set of all elements x in the set A with the property that x is related to a by the relation R.

A few examples can help make this idea clearer.

- Let A be the set of triangles in the plane and let $\sim$ be the is-similar-to relation. If T is a specific equilateral triangle, then $[\![T]\!]_\sim$ is the set of all equilateral triangles.

- Let A be the set of all people and let R be the has-the-same-birthday-as (month and day only) relation. Then $[\![\text{you}]\!]_R$ is the set of all people that have the same birthday as you. In this example, there are exactly 366 different equivalence classes. Every person is in just one of these non-overlapping sets.

- Let A be the set of lines in a plane and let $\|$ be the is-parallel-to relation. Then $[\![\ell]\!]_\|$ is the set of all lines parallel to ℓ (including ℓ itself). The set of all lines in the plane is divided into non-overlapping sets, one for each possible slope.

An element of an equivalence class is called a *representative* of that class. Here are some examples of representatives drawn from the examples above.

- Let A be the set of triangles in the plane and let $\sim$ be the is-similar-to relation. The triangle with vertices at coordinates $(0, 0)$, $(3, 0)$, and $(0, 4)$ is a representative of the equivalence class that contains all right triangles whose sides are in a 3:4:5 ratio.

- Let A be the set of all people and let R be the has-the-same-birthday-as (month and day only) relation. The actor Tom Hanks was born on July 9, so he is a representative of the equivalence class that consists of all people with that birthday.

- Let A be the set of lines in a plane and let $\|$ be the is-parallel-to relation. Then the x-axis is a representative of the set of all horizontal (slope zero) lines.

Partitions

In each of the examples just presented – indeed, for any equivalence relation – the equivalence classes form non-overlapping subsets of A with each element of a in exactly one of those subsets. More formally, equivalence classes form a *partition* of A. Here's the definition.

Definition 1.5. Let A be a set. A *partition* of A is a set whose elements (1) are subsets of A, (2) are nonempty, (3) are pairwise disjoint, and (4) have a union that is A.

We present an example and then examine the definition piece by piece. Let $A = \{1, 2, 3, 4, 5, 6, 7, 8, 9, 10\}$. The following is a partition of A:

$$\mathbf{P} = \Big\{\{1, 3, 5\}, \{2, 4, 6, 8\}, \{7\}, \{9, 10\}\Big\}.$$

Let's check that $\mathbf{P}$ is a partition of $A = \{1, 2, \ldots, 10\}$ by seeing that it satisfies the four conditions.

(1) $\mathbf{P}$ is a set and its elements are subsets of A, for example, $\{1, 3, 5\} \in \mathbf{P}$.

(2) The four sets that are elements of $\mathbf{P}$ are all nonempty.

(3) The four sets that are elements of $\mathbf{P}$ are *pairwise disjoint*. This simply means that the intersection of any two of them is the empty set:

$$\{1, 3, 5\} \cap \{2, 4, 6, 8\} = \varnothing, \quad \{1, 3, 5\} \cap \{7\} = \varnothing,$$
$$\{1, 3, 5\} \cap \{9, 10\} = \varnothing, \quad \{2, 4, 6, 8\} \cap \{7\} = \varnothing,$$
$$\{2, 4, 6, 8\} \cap \{9, 10\} = \varnothing, \quad \{7\} \cap \{9, 10\} = \varnothing.$$

(4) The union of the four sets in $\mathbf{P}$ is the original set A:

$$\{1, 3, 5\} \cup \{2, 4, 6, 8\} \cup \{7\} \cup \{9, 10\} = \{1, 2, 3, 4, 5, 6, 7, 8, 9, 10\}.$$

Figure 1.3 provides a way to visualize the partition $\mathbf{P}$.

The sets contained in a partition are called the *parts* of the partition.

Recap

In this chapter we covered the following key topics:

- Sets: unordered, repetition-free collections.
- The empty set and singleton sets.
- The set operations union and intersection.
- Subsets.
- Lists.

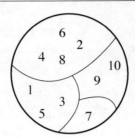

Figure 1.3: Illustration of a partition of the set $\{1, 2, \ldots, 10\}$. The partition is $\mathbf{P} = \Big\{ \{1, 3, 5\}, \{2, 4, 6, 8\}, \{7\}, \{9, 10\} \Big\}$.

- Relations.
- One-to-one correspondences.
- Equivalence relations (which are reflexive, symmetric, and transitive).
- Equivalence classes and partitions.

Exercises

1.1   a) Are $\{1, 2, 3\}$ and $\{2, 1, 3\}$ the same set?

       b) Are $(1, 2, 3)$ and $(2, 1, 3)$ the same list?

       c) Are $\{1, 1, 2, 3\}$ and $\{1, 2, 3\}$ the same set?

       d) Are $(1, 1, 2, 3)$ and $(1, 2, 3)$ the same list?

1.2 Write out all the subsets of $\{1, 2, 3\}$.

1.3 Give an example of sets A and B for which $A \not\subseteq B$ (A is not a subset of B).

1.4 Let A be a set. Explain why $\emptyset \subseteq A$.

1.5 The empty set has no elements. Does it have any subsets?

1.6 Let A and B be sets. Suppose $A \subseteq B$ and $B \subseteq A$. What does this tell you about A and B?

1.7 Let A and B be sets. It may be the case that $A \cap B = A$. When does this happen?

1.8 Find three sets A, B, and C for which $A \cap B \cap C = \emptyset$, but no pair of these sets are disjoint.

1.9 Find a set A for which both of the following are both true: $\{2\} \subseteq A$ and $\{2\} \in A$.

1.10 We define a *singleton* set without using the number 1; see Definition 1.1. In a similar vein, create a definition of a *doubleton* set – a set with exactly two members – without using the number 2.

1.11 Let A be the ten-element set $\{1, 2, 3, \ldots, 10\}$. How many ordered pairs (x, y) can be formed where x and y are both elements of A?

1.12 Let A and B be sets. Recall that $A - B$ is the set containing all elements of A that are not in B.

For example, with $A = \{1, 2, 3, 4, 5\}$ and $B = \{4, 5, 6, 7\}$ we have $A - B = \{1, 2, 3\}$.

   a) For these specific sets, what is $B - A$?

   b) For any two sets A and B, when is it the case that $A - B = B - A$?

1.13 Let A and B be sets. Their *symmetric difference*, $A \triangle B$, is the set of all elements that are in A or in B, but not in both. In symbols: $A \triangle B = (A \cup B) - (A \cap B)$.

   a) Evaluate: $\{1, 2, 3, 4\} \triangle \{3, 4, 5, 6\}$.

   b) If A is a set, what is $A \triangle A$?

   c) If A is a set, what is $A \triangle \emptyset$?

   d) If $A \subseteq B$, what is $A \triangle B$?

1.14 Given two (finite) lists, L_1 and L_2, define an operation $\oplus$ (called *concatenation*) that creates a new list containing the elements of L_1 followed by the elements of L_2. For example, $(1, 1, 3, 2) \oplus (2, 3, 0) = (1, 1, 3, 2, 2, 3, 0)$.

   a) Is this operation commutative? That is, does $L_1 \oplus L_2 = L_2 \oplus L_1$ for any two lists?

   b) Is this operation associative? That is, does $(L_1 \oplus L_2) \oplus L_3 = L_1 \oplus (L_2 \oplus L3)$ for any three lists?

   c) Is there an identity element for this operation; that is, is there a list I with the property that $L \oplus I = I \oplus L$ for any list L?

1.15 The simplest way to specify a set is to list its elements between curly braces, like this: $\{1, 3, 5\}$. It is also useful to use three dots to show a range of values, like this: $\{1, 2, 3, \ldots, 100\}$.

An alternative is *set builder notation*. The general form of this notation looks like this:
$$\{x \in S : \text{conditions on } x\}.$$
The first part, $x \in S$, tells us that we are drawing elements from the set S. The second part, conditions on x, tells us which elements to include. For example:
$$\{x \in \mathbb{N} : 0 < x \leq 100\}$$
is the set $\{1, 2, \ldots, 100\}$. If the context is clear, the notation is sometimes abbreviated to $\{x : \text{conditions on } x\}$.

The following sets are presented here either as a list or in set builder notation. In each case, express the same set in the other manner.

a) $\{x \in \mathbb{N} : x > 99\}$
b) $\{0, 2, 4, 6, 8, 10, \ldots\}$
c) $\{x \in \mathbb{N} : x > 5 \text{ and } x < 4\}$
d) $\{0, 1, 2, 3\} \cup \{5, 6, 7, 8, \ldots\}$

1.16 Let R be a relation from a set A to a set B. Define a new relation S from B to A as follows: For $a \in A$ and $b \in B$, we have $b\, S\, a$ exactly when $a\, R\, b$. We call S the *inverse* of relation R and use the following notation: $S = R^{-1}$.

a) Explain how to modify the picture of R (see Figure 1.2) to form the picture of S.
b) What is $\left(R^{-1}\right)^{-1}$?
c) If R is a one-to-one correspondence from A to B, what can we say about $S = R^{-1}$?

1.17 In this chapter we present relations as predicates. That is, given sets A and B, a relation R from A to B is expressed as a statement of the form $a\, R\, b$ where $a \in A$ and $b \in B$. Sometimes the statement is true (e.g., $3 < 10$) and sometimes it is false (e.g., $5 = 17$).

In this exercise we develop an alternative approach. Given sets A and B, the *Cartesian product* of A and B is the set of all possible ordered pairs of the form (a, b) where $a \in A$ and $b \in B$. The symbol $\times$ is used to denote the Cartesian product. Thus, in symbols we have:
$$A \times B = \{(a, b) : a \in A \text{ and } b \in B\}.$$

a) Write out the entire set $\{1, 2, 3\} \times \{4, 5\}$.
b) Is $\times$ for sets commutative? That is, does $A \times B = B \times A$?

Now we present the formal definition:

A *relation* from a set A to a set B is a subset of $A \times B$.

This is rather cryptic. On the one hand, we think of R as a statement of the form $a\ R\ b$. We can view R simply as a collection of all the pairs (a, b) for which $a\ R\ b$ is true.

For example, if $A = \{1, 2, 3\}$ and $B = \{4, 5\}$ a possible relation is

$$R = \{(1, 4), (1, 5), (2, 5), (3, 4)\}.$$

For this relation it is true that $1\ R\ 4$ and $2\ R\ 5$, but it is false that $2\ R\ 4$. Writing R as a set simply records all the pairs (a, b) for which $a\ R\ b$ is true.

    c) Write the less-than relation, $<$, for the set $\{1, 2, 3, 4\}$ to itself as a set of ordered pairs.

    d) Draw a picture of the relation from part c.

1.18 Find all one-to-one correspondences from the set $A = \{1, 2, 3\}$ to the set $B = \{4, 5, 6\}$.

1.19   a) Give two examples of human relations: one symmetric and the other not symmetric.

    b) Give two examples of common mathematical relations: one symmetric and the other not symmetric.

    c) Show how to express the fact that a relation R is symmetric using the notation from Exercise 1.16.

1.20 Let A be a set. Must there be a one-to-one correspondence from A to itself?

1.21 Let R be a one-to-one correspondence from A to B. Must there be a one-to-one correspondence from B to A?

1.22 Let A, B, and C be sets. Suppose R is a one-to-one correspondence from A to B and S is a one-to-one correspondence from B to C. Must there be a one-to-one correspondence from A to C?

1.23 For each of the following sets and relations, determine which of the three properties (reflexive, symmetric, transitive) hold, and if the relation is an equivalence relation.

    a) A is any set and R is the relation is-equal-to, $=$.

    b) A is the set of triangles in the plane and R is the relation is-congruent-to, $\cong$.

c) A is the set of natural numbers, $\{0, 1, 2, 3, \ldots\}$ and R is the relation less-than-or-equal-to, $\leq$.

d) A is the set of human beings and R is the relation is-a-sibling-of. [Note: Two different people are siblings if they have at least one parent in common, so this includes half-brothers and half-sisters.]

e) A is the set of human beings and R is the relation has-the-same-biological-mother-as.

1.24 Find all partitions of the set $\{1, 2, 3, 4\}$. Hint: There are fifteen of them.

1.25 Let A be the set of all points in the plane expressed in coordinates. That is, $A = \{(x, y) : x, y \in \mathbb{R}\}$. Define a relation R by $(a, b) \ R \ (c, d)$ provided $a^2 + b^2 = c^2 + d^2$.

  a) Show that R is an equivalence relation.

  b) Describe R's equivalence classes. In particular, describe $[\![(a, b)]\!]_R$.

1.26 Let T be the set of all triangles in the plane and let t be the triangle whose corners are at coordinates $(0, 0)$, $(1, 0)$, and $(0, 1)$. Describe the equivalence class $[\![t]\!]_\sim$ where $\sim$ is the relation is-similar-to for the set T.

1.27 Let R be a relation on the integers defined by $x \ R \ y$ provided $x - y$ is even.

  a) Is R reflexive, symmetric, and transitive? That is, is R an equivalence relation?

  b) If so, what are the equivalence classes?

1.28 Repeat Exercise 1.27 but with a modified definition of R in which $x \ R \ y$ provided $x - y$ is odd.

1.29 A *function* from set A to set B is a relation that satisfies the following properties:

  - For every $a \in A$ there is a $b \in B$ such that $a \ f \ b$.
  - If $a \ f \ b$ and $a \ f \ b'$ then $b = b'$.

  The notation $a \ f \ b$ is not the usual way to express functions; the standard[2] notation is $f(a) = b$.

  a) If we draw a picture of a function relation f as arrows from elements of A to elements of B, what do the two conditions in the definition imply about the diagram?

---

[2] An alternative notation for $f(a) = b$ is $a \mapsto b$. This notation presumes we know the function, f, under consideration.

b) If f is a function from set A to set B, then f^{-1} is a relation from B to A (see Exercise 1.16). When is f^{-1} also a function?

1.30 Given a set A we say that $\star$ is an *operation* on A if, for any two elements $a, b \in A$, $a \star b$ is also an element of A. For example, $+$ is an operation for the natural numbers. This definition is imprecise. Here is a rigorous way to define an operation.

Let A be a set. An *operation* is a function from $A \times A$ (see Exercise 1.17) to A. In words, $\star$ is a function that takes, as input, a pair of elements of A and returns, as output, an element of A. Rather than using the usual function notation $\star(a, b)$, we prefer to place the symbol between the elements like this: $a \star b$.

a) For integers a and b, suppose $a \star b = a^2 - b^2$. What is $5 \star 3$?

b) Is $-$ an operation on the natural numbers, $\mathbb{N} = \{0, 1, 2, \ldots\}$?

c) How many different operations can be defined on the set $\{1, 2\}$?

1.31 Let $\star$ be an operation on a set A. We say that $e \in A$ is an *identity element* for $\star$ provided $a \star e = e \star a = a$ for any element $a \in A$.

Can an operation have two different identity elements?

Chapter 2

$\mathbb{N}$: Natural Numbers

The goal of this book is to develop a thorough understanding of the real numbers. We begin the journey with the familiar *natural numbers* and build concept upon concept from there.

The *natural numbers* are the answers to counting problems. We use a stylized capital N to denote the set of natural numbers:

$$\mathbb{N} = \{0, 1, 2, 3, \ldots\}.$$

2.1 You Can Skip this Chapter, but Don't

The natural numbers are so familiar that it seems overly fussy to try to define them. Most mathematicians are happy to take the natural numbers as given and proceed to more interesting matters.

This is the essence of the philosophy espoused by the nineteenth-century German mathematician Leopold Kronecker. He is purported to have said:

> *Die ganzen Zahlen hat der liebe Gott gemacht, alles andere ist Menschenwerk.*
>
> *God made the integers, all else is the work of man.*

The use of theological and gendered language isn't necessary to make the point that we can just take integers (or, more simply, the natural numbers) as a starting point and work from there.

In other words, let's not worry about trying to figure out what the integers (natural numbers and their negatives) are or where they come from; let's just get to work using them and move on. His attitude is one embraced by many mathematicians in their daily work. We don't spend much time pondering the foundational framework for these most basic numbers.

> **No Consensus on Zero**
>
> While the concepts of integers, rational numbers, and real numbers are universally agreed upon in the mathematics community, there is a lack of consensus for natural numbers. The disagreement centers around the number zero. For some mathematicians (myself included) it is best to include 0 as a natural number. Others prefer to omit 0 from $\mathbb{N}$. Curiously, different standards organizations give divergent recommendations; see [3], which does not include 0, versus [4], which does.

Here's a rough analogy: Architects want to design beautiful, functional buildings. While they need to be concerned with the soil and bedrock on which their structures sit, those issues are best handled by geotechnical engineers.

While we could just start our journey to the real number system by taking the natural numbers as given, we start with something more basic. Besides getting closer to the bedrock, the approach we take in this chapter is recapitulated in many of the following chapters. The recipe we follow is (a) start with something we know, (b) make an equivalence relation, and then (c) form equivalence classes. That's likely incomprehensible on a first reading, so let's move on. We need to start somewhere, although it seems we have nothing to start with.

Our approach is rooted in the work of Gottlob Frege, a nineteenth-century German mathematician and philosopher. See, for example, [1].

2.2 Starting from Nothing

Mathematics is teeming with all kinds of numbers from the simple 1, 2, 3 to negative numbers such as -5, to fractions such as $\frac{2}{3}$, to the more esoteric such as π, and the imaginary such as i. Our objective is to present a solid understanding of the concept of a real number. Our approach is to build one type of number based on an earlier, simpler type of number, starting with the natural numbers.

Ironically, beginning with the "easy" natural numbers presents a daunting challenge: How do we define the natural numbers? From what earlier idea(s) do they arise? There are no numbers more basic than the natural numbers on which to build. And supposing we have a concept more basic than natural numbers, from what even more basic concept(s) do they arise? Perhaps it's "turtles all the way down"? See [16].

There are various solutions to this conundrum.

Logicians are mathematicians who study the foundations of mathematics. With Herculean effort, they developed a path to define natural numbers starting from "nothing." All of mathematics can be defined from sets, so logicians assume the existence of a specific set, namely the empty set $\emptyset$. In this sense, they start from "nothing" to build a universe of sets and numbers. This is not an easy road to follow. Perhaps some day you would like to explore this approach, but we do not fully embrace it in this book.

In this chapter we take an intermediate approach. We are not going to descend all the way to bottom of the foundational rabbit hole. Nor are we simply going to ignore the problem of trying to define the natural numbers.

We use finite sets to define the natural numbers, the addition and multiplication operations, and the less-than-or-equal relation: $(\mathbb{N}, +, \cdot, \leq)$.

2.3 Stuff to Count

Natural numbers are the answers to counting problems – questions of the form "How many?" Let's pause a moment to consider: What are we counting?

There are many possible examples of counting from the physical world: How many shoes do I have in my closet? How many people live in my town? How many words are in this book? How many atoms are in a gram of carbon?

However, our goal is to create a tidy mathematical development of counting to underly the notion of natural numbers. That means we need things to count.

To this end, we introduce an abstract notion of *basic objects*.

One way we might do this is to simply say that there is, in the universe of mathematics, an unlimited supply of basic stuff – objects we can count. So if we have a set A, we can always make a bigger set by adding something to A. What are we including? Some bit of basic stuff.

For technical reasons, we're going to be more explicit about what we mean by basic stuff. To begin, we just assume that there are distinct objects a and b. We don't say what they are. They just are.

We then define additional basic objects to be any pair (x, y) where both x and y are themselves basic objects.

Definition 2.1. An object Z is a *basic object* if one of the following is true:

- $Z = a$,
- $Z = b$,
- $Z = (x, y)$ where x and y are both basic objects.

Let's see where this leads. We have a and b as basic objects. Then we also have

$$(a, a), \quad (a, b), \quad (b, a), \quad \text{and} \quad (b, b)$$

as additional basic objects. We're not done! There is a lot more basic stuff we can make by pairing any of the six basic objects we've already found. All of the following are basic objects that are distinct from the ones we have already identified:

$(a,(a,a))$, $\quad (a,(a,b))$, $\quad (a,(b,a))$, $\quad (a,(b,b))$,
$((a,a),a)$, $\quad ((a,b),a)$, $\quad ((b,a),a)$, $\quad ((b,b),a)$,
$(b,(a,a))$, $\quad (b,(a,b))$, $\quad (b,(b,a))$, $\quad (b,(b,b))$,
$((a,a),b)$, $\quad ((a,b),b)$, $\quad ((b,a),b)$, $\quad ((b,b),b)$,

$((a,a),(a,a))$, $\quad ((a,a),(a,b))$, $\quad ((a,a),(b,a))$, $\quad ((a,a),(b,b))$,
$((a,b),(a,a))$, $\quad ((a,b),(a,b))$, $\quad ((a,b),(b,a))$, $\quad ((a,b),(b,b))$,
$((b,a),(a,a))$, $\quad ((b,a),(a,b))$, $\quad ((b,a),(b,a))$, $\quad ((b,a),(b,b))$,
$((b,b),(a,a))$, $\quad ((b,b),(a,b))$, $\quad ((b,b),(b,a))$, $\quad ((b,b),(b,b))$.

We can use those (in addition to the first six) to make even more basic stuff.

2.4 Prelude to Counting: Sets of the Same Size

Chapter 1 introduces ideas that we use here to create the natural numbers. The key ideas we need are these:

- sets and their operations (Section 1.1),
- one-to-one correspondences (Section 1.3),
- equivalence relations and equivalence classes (Section 1.4).

Our approach is to show that natural numbers are the "answers" to counting problems.

One-to-one correspondences are central to understanding the idea of counting. For example, the set of letters in the Latin alphabet is $\{a, b, c, \ldots, z\}$. There is a one-to-one correspondence between the alphabet and the numbers 1 through 26:

$$a \to 1, \ b \to 2, \ c \to 3, \ \ldots, \ z \to 26.$$

If X is another set of 26 things, there is a one-to-one correspondence between X and the numbers 1 through 26. Of course, we have yet to define natural numbers, so we have gotten ahead of ourselves.

Here's the interesting thing. Suppose X and Y are two different sets, both with the same number of elements – say, 26 in each. We do not need to use numbers at all to show that they have the same size! All we need to do is find a one-to-one correspondence from X to Y.

This can be confusing: How can we say two sets contain the same number of elements if we don't have numbers defined yet? The answer is simply this: Given two sets, we say they *have the same size* if there is a one-to-one correspondence between them. Now we could use a technical term for this (and avoid the word "size"); we could call two such sets *equinumerous*, but that's just being fancy. We call this relation *has-a-one-to-one-correspondence-with*.

> **The Shifting Meaning of ≡**
>
> The symbol ≡ appears in this chapter and recurs in subsequent chapters, but with different meanings. Within each chapter, the symbol ≡ has just one meaning that lasts for the entire chapter. In subsequent chapters, its meaning changes but, in every case, ≡ is an equivalence relation. For better or worse, mathematicians often allow a single symbol to have different meanings depending on context; the alternative would be to have an unimaginably large palette of symbols.
>
> In this chapter we use the symbol ≡ to stand for the has-a-one-to-one-correspondence-with relation.

It helps to introduce a bit of notation. Suppose A and B are sets. We write $A \equiv B$ to mean that there is a one-to-one correspondence from A to B. In words, $A \equiv B$ means the set A has a one-to-one correspondence with the set B.

Notice that ≡ is, itself, a relation between sets. What sort of relation is it? Exercises 1.20–22 show that ≡ is an equivalence relation.

Let's start with an imprecise introduction to natural numbers. Imagine that A and B are singleton sets (see Definition 1.1 on page 8). For instance, $A = \{a\}$ and $B = \{b\}$ where a and b are some objects. Then quite obviously $A \equiv B$; that is, A has a one-to-one correspondence with B in which $a \rightarrow b$. All singleton sets are ≡-equivalent to each other.

Suppose X is any set that is not a singleton, then X either is empty or contains at least two distinct elements. In that case, X is not ≡-equivalent to a singleton set.

Thus, not only are singleton sets all ≡-equivalent to each other, but no other sets are ≡-equivalent to singleton sets.

The natural number **1** is the "property" that all singleton sets share. Now this is rather vague, but gets at the key idea. A set has size **1** means it has a one-to-one correspondence with a singleton set.

What is **0**? It is the "property" shared by all empty sets. Of course, there is only one empty set, ∅. No other set has a one-to-one correspondence with ∅. To say that a set has size **0** means it has a one-to-one correspondence with the empty set.

Each natural number is the "unique property" shared by finite sets that are all ≡-equivalent to each other. We are going to make this more precise, but we have a problem: Intuitively, a *finite* set is one with only finitely many elements. Alas, we have used the concept of *finite* to define *finite* and that appears to be a crime against logic. We might say that a set is finite if the number of elements in the set is a natural number (so, for example, ℕ is not finite), but our plan is to

> **What's with the Boldface Numbers?**
>
> In this and subsequent chapters we introduce new types of numbers. We use boldface numerals and variables to stand for the new types of numbers introduced in each chapter. Hence, in this chapter, natural numbers are shown in boldface type.
> Next, in Chapter 3, we define the integers in terms of the natural numbers. There we use boldface to stand for the integers and "demote" natural numbers to a lightweight typeface. In Chapter 5 we introduce rational numbers and in that chapter boldface stands for the new type of number we are introducing and integers are relegated to the light typeface.

use the concept of a finite set to define natural numbers, and not the other way around.

2.5 Finite Sets

We have an intuitive sense of *finite* versus *infinite* sets. The infinite ones "go on forever" while the finite ones don't. Unfortunately, we don't have a rigorous concept of "go on forever." If we had the natural numbers at the ready, we could say that a set A is finite if it has a one-to-one correspondence to a set of the form $\{1, 2, \ldots, n\}$ (when $n = 0$, this is $\varnothing$). But this is backwards; our plan is to use finite sets to define the natural numbers, so we need another approach.

Let's start with something we know intuitively. The empty set $\varnothing$ is a finite set. It has no elements; nothing can be further from the infinite! Also, singleton sets are finite. We use these as the basis for the following definition proposed by the twentieth-century Polish mathematician Kazimierz Kuratowski:

> **Definition 2.2.** A set is *finite* if it is the empty set or the union of a finite set and a singleton set.

At first glance this is a troubling definition because it defines the word *finite* in terms of itself. However, this works because it is a recursive definition.

For example, consider the set $\{a, b, c\}$ (where a, b, and c are distinct objects). Is it, by Definition 2.2, finite?

First note that $\{a\}$ is finite because, by the definition, it is the union of the empty set (finite by definition) and a singleton (itself): $\{a\} = \varnothing \cup \{a\}$. Since $\{b\}$ is a singleton and we know $\{a\}$ is finite, then $\{a, b\}$ is finite because

$\{a, b\} = \{a\} \cup \{b\}$. Finally, because $\{a, b\}$ is finite and $\{c\}$ is a singleton, their union $\{a, b\} \cup \{c\} = \{a, b, c\}$ is finite.

In other words, finite sets are created by starting with the empty set and appending singleton elements one by one.

Combining pairs of finite sets using the usual operations of set theory yields finite sets.

Proposition 2.3. *For finite sets A and B we have:*

- *$A \cup B$ is finite,*
- *$A \cap B$ is finite,*
- *$A \times B$ is finite.*

Let's see why this is correct.

- $A \cup B$ is finite: We are given that A is finite and we know that B is built up from $\emptyset$ by appending singletons. Simply append those singletons to A (appending a singleton to a finite set gives a finite set) and the result is $A \cup B$.

- $A \cap B$ is finite: We form A from $\emptyset$ by appending singletons. Instead of appending all the elements of A, append only those elements that are also in B. The result is $A \cap B$.

- $A \times B$ is finite: Recall that $A \times B = \{(a, b) : a \in A \text{ and } b \in B\}$. (See Exercise 1.17.)

   Let's start with a simpler case: $A \times \{y\}$ is finite. Since A is finite, it is constructed by appending singleton elements a one at at time starting from $\emptyset$. Now, instead of adding elements a one at a time to $\emptyset$, add the pairs (a, y) one at a time. The result is $A \times \{y\}$, which therefore is finite.

   More generally, let's consider $A \times B$. We just explained that for each $b \in B$ the set $A \times \{b\}$ is finite. The full set $A \times B$ is formed by the step-by-step union of a finite set (starting with $\emptyset$) and a finite set of the form $A \times \{b\}$. Since the union of finite sets is finite, we find that $A \times B$ is finite.

2.6 Natural Numbers are Equivalence Classes

We now have the necessary tools to define natural numbers. Earlier we vaguely described the natural numbers as the "answers to counting problems." We can now be much more precise. We use the following ingredients:

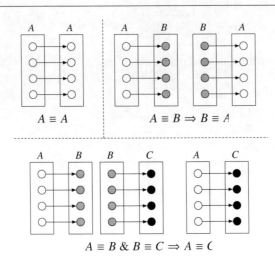

Figure 2.1: This figure illustrates that the relation has-a-one-to-one-correspondence-with for finite sets is an equivalence relation. The top left portion illustrates that $\equiv$ is reflexive, the top right portion illustrates that $\equiv$ is symmetric, and the bottom portion illustrates that $\equiv$ is transitive.

- Finite sets of basic objects.
- The has-a-one-to-one-correspondence-with relation, $\equiv$. This is an equivalence relation as shown by Exercises 1.20–22. See also Figure 2.1.
- Equivalence classes of $\equiv$.

Recall that if A is a set, then $[\![A]\!]_\equiv$ are all the sets X with $A \equiv X$. (See Definition 1.4.) Since $\equiv$ is the only equivalence relation considered in this chapter, we omit the subscript and write $[\![A]\!]$.

Definition 2.4. A *natural number* is the $\equiv$-equivalence class of a finite set of basic objects.

For example, the number **0** is simply $[\![\varnothing]\!]$. There is only one empty set so $[\![\varnothing]\!]$ contains only $\varnothing$.

The number **1** is more interesting. It is the equivalence class of all singleton sets of basic objects; that is, $\mathbf{1} = [\![\{t\}]\!]$ where t is any basic object.

> **It's all just Counting**
>
> Scratch below the surface of all the mathematical jargon and notation and you'll see that everything about the natural numbers is simply counting.
>
> Why is 12 + 5 = 5 + 12? Suppose we have two sacks of marbles: one with 12 marbles and the other with 5. How many marbles do we have all together? We can count, starting with the sack of 12 and then continuing with the sack of 5, or we could reverse the order of the sacks. It doesn't matter; we'll get 17 either way.
>
> Why is 12 · 5 = 5 · 12? As repeated addition, why are these equal:
>
> $$\underbrace{12 + 12 + 12 + 12 + 12}_{5 \text{ terms}} = \underbrace{5 + 5 + 5 + 5 + 5 + 5 + 5 + 5 + 5 + 5 + 5 + 5}_{12 \text{ terms}}?$$
>
> Just imagine a 12-by-5 checkerboard and count the squares. If we count row-by-row, we are adding twelve 5s. If we count column-by-column, there are five 12s. Both are correct answers to the same question, so they must be equal.
>
> Why is 5 < 12? Back to the sacks of marbles. If we pull one marble each from the two sacks, and keep repeating, the 5-marble sack empties first.

What is **2**? Suppose s and t are different basic objects. Note that $\{s,t\}$ is a finite set because it is the union of the singletons $\{s\}$ and $\{t\}$. The number **2** is the equivalence class of all sets of basic objects that have a one-to-one-correspondence with $\{s,t\}$; in other words $\mathbf{2} = [\![\{s,t\}]\!]$. See Figure 2.2.

We have used the ideas of equivalence relations, equivalence classes, and finite sets to make the vague idea that a natural number, say **2**, is a "property" shared by all two-element sets, into the more precise (if elaborate) $\mathbf{2} = [\![\{s,t\}]\!]$.

The set of natural numbers, $\mathbb{N}$, is simply the collection of all the equivalence classes of the form $[\![A]\!]$ where A is a finite set of basic objects. To complete the story we need to ascribe meanings to the symbols +, ·, and ≤.

2.7 Addition

Let **a** and **b** be natural numbers. We wish to define **a** + **b**. Let's start with the incredibly simple **1** + **1**. Of course, we expect the answer to be **2**.

Recall that **1** is the equivalence class of singleton sets and **2** is the equivalence class $[\![\{s,t\}]\!]$ where s and t are distinct basic objects. Note that $\{s\}$ is a singleton set and therefore is a representative of **1**. (Recall that this simply means that $\{s\}$ is a member of the class **1**. See the discussion on page 14.)

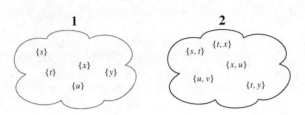

Figure 2.2: The natural number **1** is the equivalence class of all singleton sets of basic objects. The natural number **2** is the equivalence class of all sets of the form $\{p,q\}$ where p and q are distinct basic objects. In this illustration, objects s, t, u, v, x, and y are all distinct basic objects.

The singleton $\{t\}$ is also a representative of **1** and the set $\{s,t\}$ is a representative of the class **2**.

Addition is (almost) this simple: To add **a** and **b**, take a representative A from **a**, a representative B from **b**, form their union $A \cup B$, and the sum is simply $[\![A \cup B]\!]$.

This is nearly correct, but we need to be careful. Let's add **1** + **2**. The singleton set $\{s\}$ is a representative set of **1** and the set $\{s,t\}$ (with $s \ne t$) is a representative of **2**. However, their union $\{s\} \cup \{s,t\}$ is not a three-element set; it's just $\{s,t\}$.

To add **1** and **2** we need to pick representatives that are disjoint. Let's try again. Choose $\{s\}$ as a representative of **1** and choose $\{t,u\}$ as a representative of **2** with the proviso that $\{s\}$ and $\{t,u\}$ are disjoint (i.e., $\{s\} \cap \{t,u\} = \varnothing$).

The union of these two sets is $\{s\} \cup \{t,u\} = \{s,t,u\}$. The result is a finite set, but it is neither in the class **1** nor in the class **2**. It's in an entirely new class that we call, for obvious reasons, **3**. See Figure 2.3.

Definition 2.5 (Addition of natural numbers). Let **a** and **b** be natural numbers. Define the sum **a** + **b** to be $[\![A \cup B]\!]$ where A is a representative of **a**, B is a representative of **b**, and $A \cap B = \varnothing$.

We say that A and B are *disjoint* representatives of **a** and **b**, respectively.

An important detail needs to be checked to be sure this definition is correct. Suppose A and B are disjoint representatives of **a** and **b**, respectively. Let's say that A' and B' are another pair of disjoint representatives of **a** and **b**. To be sure that **a** + **b** is unambiguously defined, it had better be the case that, whether we

2 N: Natural Numbers   33

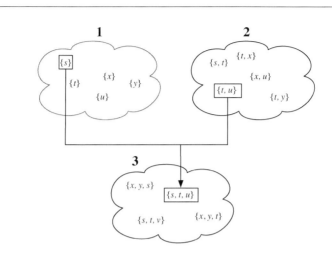

Figure 2.3: An illustration of illustration of **1** + **2** = **3**. The set $\{s\}$ is a representative of **1** and the set $\{t, u\}$ is a representative of **2**. Since s, t, and u are distinct basic objects, we have $\{s\} \cap \{t, u\} = \emptyset$. Therefore $\{s\} \cup \{t, u\} = \{s, t, u\}$ is a representative of **1** + **2**.

use A and B as the representatives or we use A' and B', we get the same answer. That is, are $[\![A \cup B]\!]$ and $[\![A' \cup B']\!]$ the same? Let's check that this is the case.

Both A and A' are in the equivalence class **a**. That means that $A \equiv A'$, so there is a one-to-one correspondence R between A and A'. Likewise, there is a one-to-one correspondence S between B and B'. When we combine these correspondences the result is a one-to-one correspondence between $A \cup B$ and $A' \cup B'$ as shown in Figure 2.4.

(If we think of the relations R and S as sets of ordered pairs, the the combined relation is simply $R \cup S$. See Exercise 1.17.)

This shows that whether we define $\mathbf{a} + \mathbf{b}$ as $[\![A \cup B]\!]$ or as $[\![A' \cup B']\!]$, the result is the same because $A \cup B \equiv A' \cup B'$ and therefore $[\![A \cup B]\!] = [\![A' \cup B']\!]$.

With this definition we can verify the basic properties of addition. For example, $\mathbf{a} + \mathbf{b} = \mathbf{b} + \mathbf{a}$; here's why. Choose disjoint representatives A and B of **a** and **b**, respectively. Since $A \cup B = B \cup A$, we have

$$\mathbf{a} + \mathbf{b} = [\![A \cup B]\!] = [\![B \cup A]\!] = \mathbf{b} + \mathbf{a}.$$

It's also easy to see that **0** is the identity element for addition. To calculate $\mathbf{a} + \mathbf{0}$ pick a representative A of **a** and note

$$\mathbf{a} + \mathbf{0} = [\![A \cup \emptyset]\!] = [\![A]\!] = \mathbf{a}.$$

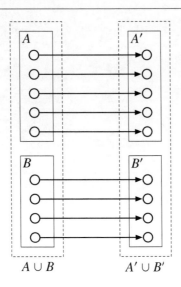

Figure 2.4: This figure illustrates that addition of natural numbers is well defined. Here A and B are disjoint sets, A' and B' are disjoint sets, $A \equiv A'$, and $B \equiv B'$. Combining the one-to-one correspondence of A and A' with the one-to-one correspondence of B and B' gives a one-to-one correspondence between their respective unions and therefore $A \cup B \equiv A' \cup B'$. This gives $[\![A \cup B]\!] = [\![A' \cup B']\!]$.

The properties of the addition operation for the natural numbers are summarized here:

(A1) If **a** and **b** are natural numbers, then **a** + **b** is also in $\mathbb{N}$. [Closure property]

(A2) If **a**, **b** ∈ $\mathbb{N}$, then **a** + **b** = **b** + **a**. [Commutative property]

(A3) If **a**, **b**, **c** ∈ $\mathbb{N}$, then (**a** + **b**) + **c** = **a** + (**b** + **c**). [Associative property]

(A4) There is an element **0** ∈ $\mathbb{N}$ with the property that for all **a** ∈ $\mathbb{N}$ we have **a** + **0** = **a**. [Identity element]

(A5) If **a**, **b**, and **x** are natural numbers with **a** + **x** = **b** + **x**, then **a** = **b**. [Cancellation property]

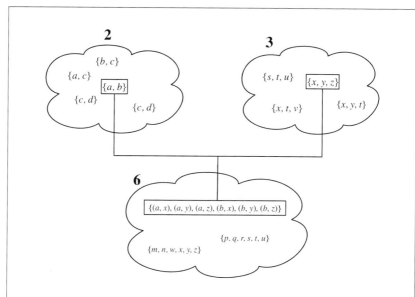

Figure 2.5: Illustrating the multiplication of **2** times **3**. Here $\{a,b\}$ is a representative of **2** and $\{x,y,z\}$ is a representative of **3**. Then

$$\{a,b\} \times \{x,y,z\} = \{(a,x), (a,y), (a,z), (b,x), (b,y), (b,z)\}$$

is a representative of $\mathbf{2} \cdot \mathbf{3}$.

2.8 Multiplication

Given natural numbers **a** and **b**, what is $\mathbf{a} \cdot \mathbf{b}$? As we did for addition, we begin with representatives A and B of **a** and **b**, respectively. Recall the *Cartesian product* of sets (see Exercise 1.17 on page 18):

$$A \times B = \{(a,b) \colon a \in A \text{ and } b \in B\}.$$

That is, $A \times B$ is the set of all ordered pairs of the form (a,b) where a is chosen from A and b is chosen from B. Then $\mathbf{a} \cdot \mathbf{b}$ is simply $[\![A \times B]\!]$. See Figure 2.5.

Definition 2.6 (Multiplication of natural numbers). Let **a** and **b** be natural numbers. Let A be a representative of **a** and let B be a representative of **b**. Then

$$\mathbf{a} \cdot \mathbf{b} = [\![A \times B]\!].$$

Note: Since A and B are sets of basic objects, so is $A \times B$; see Exercise 2.14.

As with addition, we need to pause for a moment to check if this definition is consistent. That is to say, if we have two representatives for **a**, say A and A', and two representatives for **b**, say B and B', do we have $[\![A \times B]\!] \equiv [\![A' \times B']\!]$?

The answer is yes, and here's why. We are given that $[\![A]\!] = [\![A']\!]$ and $[\![B]\!] = [\![B']\!]$, so there is a one-to-one correspondence between A and A', and between B and B'. Consider $(a, b) \in A \times B$. We know that a pairs with an a' in A' and that b pairs with b' in B. So we simply pair (a, b) with (a', b') and we have a one-to-one correspondence between $A \times B$ and $A' \times B'$. Thus $[\![A \times B]\!] = [\![A' \times B']\!]$, so the definition of multiplication is consistent.

Properties of multiplication flow nicely from this definition. Here are some examples.

- Multiplication is commutative. Let **a** and **b** be natural numbers and let A and B be representatives, respectively.

  Notice there is a simple one-to-one correspondence between $A \times B$ and $B \times A$ in which (a, b) in $A \times B$ is paired with $(b, a) \in B \times A$. This gives $A \times B \equiv B \times A$ and so $[\![A \times B]\!] = [\![B \times A]\!]$. Therefore
  $$\mathbf{a} \cdot \mathbf{b} = [\![A \times B]\!] = [\![B \times A]\!] = \mathbf{b} \cdot \mathbf{a}.$$

- Multiplication by **0**. Let **a** be a natural number and let A be a representative. Note that $A \times \varnothing = \varnothing$ and so $\mathbf{a} \cdot \mathbf{0} = [\![A \times \varnothing]\!] = [\![\varnothing]\!] = \mathbf{0}$.

- Multiplication by **1**. Use $\{s\}$ as a singleton representing **1**. Let A be a representative of a natural number **a**. Note that $A \times \{s\}$ is the set $\{(a, s): a \in A\}$. There is a simple one-to-one correspondence between $A \times \{s\}$ and A in which $(a, s) \to a$. Therefore we have
  $$\mathbf{a} \cdot \mathbf{1} = [\![A \times \{a\}]\!] = [\![A]\!] = \mathbf{a}.$$

- The distributive property. Let **a**, **b**, and **c** be natural numbers and choose representatives A, B, and C for them, respectively, with B and C disjoint. The expression $\mathbf{a} \cdot (\mathbf{b}+\mathbf{c})$ evaluates to $[\![A \times (B \cup C)]\!]$. Aided by Figure 2.6, notice that
  $$A \times (B \cup C) = (A \times B) \cup (A \times C).$$
  Since B and C are disjoint, it follows that $A \times B$ and $A \times C$ are disjoint (the second entries in the lists in $A \times B$ have nothing in common with the second entries of the members of $A \times C$). Therefore
  $$\begin{aligned} \mathbf{a} \cdot (\mathbf{b} + \mathbf{c}) &= [\![A \times (B \cup C)]\!] \\ &= [\![(A \times B) \cup (A \times C)]\!] \\ &= [\![A \times B]\!] + [\![A \times C]\!] \\ &= \mathbf{a} \cdot \mathbf{b} + \mathbf{a} \cdot \mathbf{c}. \end{aligned}$$

2   ℕ: Natural Numbers   37

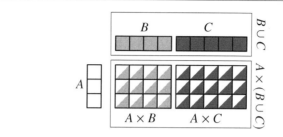

Figure 2.6: Illustrating the distributive property $a \cdot (b + c) = a \cdot b + a \cdot c$. Observe that $A \times (B \cup C) = (A \times B) \cup (A \times C)$.

Here is a summary of the properties of multiplication.

(M1) If $a, b \in \mathbb{N}$, then $a \cdot b$ is also in $\mathbb{N}$. [Closure property]

(M2) If $a, b \in \mathbb{N}$ then $a \cdot b = b \cdot a$. [Commutative property]

(M3) If $a, b, c \in \mathbb{N}$ then $(a \cdot b) \cdot c = a \cdot (b \cdot c)$. [Associative property]

(M4) There is an element $1 \in \mathbb{N}$, $1 \neq 0$, with the property that for all $a \in \mathbb{N}$ we have $a \cdot 1 = a$. [Identity element]

(D) If $a, b, c \in \mathbb{N}$ then $a \cdot (b + c) = a \cdot b + a \cdot c$. [Distributive property]

2.9   Less-than-or-equal

We defined addition of natural numbers based on union, and multiplication based on Cartesian product. In a similar way, we define $\leq$ based on subset, $\subseteq$. Imagine you have two sacks of marbles. Simultaneously you remove one marble from the first bag and one marble from the second. You repeat this over and over until one of the bags is empty. If both bags are empty, you've found a one-to-one correspondence; but if not, then the empty bag had fewer marbles in it than the other.

Translating this basic idea into mathematical language, we have this:

> **< or ≤?**
>
> We use ≤, less-than-or-equal, as the basic ordering relation. The relation < is derived from ≤ by this definition:
>
> - $a < b$ means $a \leq b$ and $a \neq b$.
>
> From there, we define ≥ and > as follows:
>
> - $a \geq b$ means $b \leq a$.
> - $a > b$ means $b \leq a$ and $b \neq a$.
>
> As best fits the situation, we use any of these four relation symbols with the understanding that <, ≥, and > are all derived from the foundational ≤. Of course, we could start with < as the basic relation and use that to define ≤, >, and ≥. See Exercise 2.17.

> **Definition 2.7** (Less-than-or-equal for natural numbers). Let **a** and **b** be natural numbers and let A and B be representatives, respectively. We say **a** is *less than or equal to* **b**, and we write $\mathbf{a} \leq \mathbf{b}$ if there is a one-to-one correspondence between A and a subset of B. See Figure 2.7.
>
> In other words, $\mathbf{a} \leq \mathbf{b}$ provided there is a $B' \subseteq B$ so that $A \equiv B'$.

The relation ≤ satisfies the following properties:

> (L1) For all $\mathbf{a} \in \mathbb{N}$, $\mathbf{a} \leq \mathbf{a}$. [Reflexive property]
>
> (L2) For all $\mathbf{a}, \mathbf{b} \in \mathbb{N}$, one (or both) of the following must be true: $\mathbf{a} \leq \mathbf{b}$ or $\mathbf{b} \leq \mathbf{a}$.
>
> (L3) For all $\mathbf{a}, \mathbf{b} \in \mathbb{N}$ if $\mathbf{a} \leq \mathbf{b}$ and $\mathbf{b} \leq \mathbf{a}$ then $\mathbf{a} = \mathbf{b}$. [Antisymmetric property]
>
> (L4) For all $\mathbf{a}, \mathbf{b}, \mathbf{c} \in \mathbb{N}$ if $\mathbf{a} \leq \mathbf{b}$ and $\mathbf{b} \leq \mathbf{c}$, then $\mathbf{a} \leq \mathbf{c}$. [Transitive property]

Just as the distributive property (D) involves the interaction of the operations

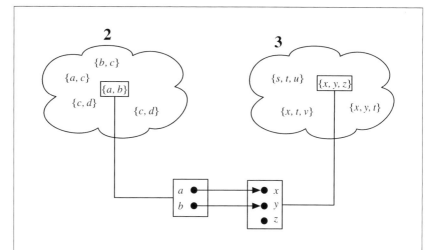

Figure 2.7: Illustrating **2** ≤ **3**. Here, $\{a,b\}$ is a representative of **2** and $\{x,y,z\}$ is a representative of **3**. The arrows show a one-to-one correspondence between $\{a,b\}$ and a subset of $\{x,y,z\}$.

of addition and multiplication, we have the following properties that describe the connection between the ≤ relation and the + and · operations:

(LA) For all **a, b, c** ∈ ℕ, if **a** ≤ **b**, then **a** + **c** ≤ **b** + **c**.

(LM) For all **a, b, c** ∈ ℕ, if **a** ≤ **b**, then **a** · **c** ≤ **b** · **c**.

Well-ordering Property

There is one additional property of ≤ that distinguishes the natural numbers ℕ from other sorts of numbers (integers, rationals, reals). To begin, the natural numbers have a least element, **0**. But more than this, one cannot have an infinite descent of natural numbers. That is, if we have a sequence of strictly decreasing natural numbers, such a list must be finite. See Exercise 2.22.

We can recast this idea without referring to finite or infinite sequences like so:

(WO) Let A be a nonempty subset of $\mathbb{N}$. Then A has a least element. [Well-ordering property]

The well-ordering property is closely linked to the concept of mathematical induction that is often used in proofs.

(MI) Let A be a subset of $\mathbb{N}$. If $\mathbf{0} \in A$ and whenever $\mathbf{k} \in A$ we have $\mathbf{k} + \mathbf{1} \in A$, then $A = \mathbb{N}$.

Proofs that employ mathematical induction are especially prevalent in discrete mathematics and number theory. See [10].

2.10 The Peano Axioms

(This section is optional.) We have defined the natural numbers as equivalence classes of finite sets. In this section, we sketch an alternative way to define the natural numbers.

The properties presented in this chapter [(A1)–(A5), (M1)–(M4), (D), (L1)–(L4), (LA), (LM), (WO)] give an exhaustive description of the natural numbers with its operations $+$ and $\cdot$ and its ordering relation $\leq$. That is, the natural numbers, and only the natural numbers, satisfy all of these properties. So, we could simply say "natural numbers are things that satisfy all these properties" and get on with our work. In other words, let's not worry about what the natural numbers are, but just assume they exist and behave in a certain way.

Mathematicians often prefer to cull the defining attributes of a concept to a bare minimum; that is, to present as few conditions as possible to specify the idea. In the case of the natural numbers, this is accomplished by a suite of axioms[1] developed by the Italian mathematician Giuseppe Peano (1858–1932).

The Peano formulation of the natural numbers consists of a set $\mathbb{N}$ and a function (see Exercise 1.29 on page 20) called next[2] that takes a natural number as input and returns a natural number as its output.

The following three assumptions are made about $\mathbb{N}$ and next:

---

[1] An *axiom* in mathematics can be thought of as either an unproven assumption or, as in this case, part of a list of defining attributes.
[2] In many treatments of the Peano axioms next is called the *successor* function.

> (P1) There is an element $\mathbf{0} \in \mathbb{N}$ with the property that there is no $\mathbf{x} \in \mathbb{N}$ with $\text{next}(\mathbf{x}) = \mathbf{0}$.
>
> (P2) If $\mathbf{a}, \mathbf{b} \in \mathbb{N}$ with $\text{next}(\mathbf{a}) = \text{next}(\mathbf{b})$, then $\mathbf{a} = \mathbf{b}$.
>
> (P3) If $A \subseteq \mathbb{N}$ has the properties
>
> - $\mathbf{0} \in A$,
> - whenever $\mathbf{n} \in A$, we also have $\text{next}(\mathbf{n}) \in A$,
>
> then $A = \mathbb{N}$.

The intuition is that $\text{next}(\mathbf{n})$ is the next larger natural number. In this scheme, we define $\mathbf{1}$ to be $\text{next}(\mathbf{0})$. If addition were defined we would see that $\text{next}(\mathbf{n})$ is simply $\mathbf{n} + \mathbf{1}$. It's not difficult to give a definition of the number $\mathbf{1}$ in this system (see Exercise 2.25); defining addition is more complicated.

Condition (P1) says that $\mathbf{0}$ is not the successor of any other natural numbers; informally, this says that $\mathbf{0}$ is the first natural number.

Condition (P2) establishes a one-to-one correspondence between the full set of natural numbers with the positive natural numbers:

$$\mathbf{0} \to \text{next}(\mathbf{0}) = \mathbf{1},$$
$$\mathbf{1} \to \text{next}(\mathbf{1}) = \mathbf{2},$$
$$\mathbf{2} \to \text{next}(\mathbf{2}) = \mathbf{3}.$$
$$\vdots$$

Finally, condition (P3) is known as *mathematical induction*. As mentioned earlier, mathematical induction is a close cousin of the well-ordering property.

The Peano axioms say nothing about the operations $+$ or $\cdot$, nor do they specify the $\leq$ relation. Rather, these are defined as follows.

Let $\mathbf{n} \in \mathbb{N}$. The operation $+$ for $\mathbb{N}$ is defined by these two rules:

- $\mathbf{n} + \mathbf{0}$ is $\mathbf{n}$,
- $\mathbf{n} + \text{next}(\mathbf{a})$ is $\text{next}(\mathbf{n} + \mathbf{a})$.

Two comments:

First, this description covers all cases because it defines $\mathbf{n} + \mathbf{x}$ for all possible values of $\mathbf{x}$. Either $\mathbf{x}$ is zero (use the first rule) or $\mathbf{x} \neq \mathbf{0}$ in which case $\mathbf{x} = \text{next}(\mathbf{a})$ for some $\mathbf{a} \in \mathbb{N}$ (use the second rule).

Second, this is an example of a *recursive* definition as the definition + refers to itself. However, it is not circular. For example, to calculate **n** + **2** we have

$$\mathbf{n} + \mathbf{2} = \mathbf{n} + \text{next}(\mathbf{1}) = \text{next}(\mathbf{n} + \mathbf{1}),$$
$$\mathbf{n} + \mathbf{1} = \mathbf{n} + \text{next}(\mathbf{0}) = \text{next}(\mathbf{n} + \mathbf{0}) = \text{next}(\mathbf{n}).$$

From this we find that **n** + **2** = next(next(**n**)).

In a similar, way we can define multiplication · (see Exercise 2.26) and ≤. From there, with a great deal of work, one can verify all of the properties of the natural number system presented earlier in this chapter.

Deriving all these properties starting only with the Peano axioms is a long, interesting story, but it is not our story. Our point is that we have an alternative to defining natural numbers as equivalence classes of finite sets. Instead, we don't even try to say what natural numbers are, but simply specify them by their properties. The three Peano axioms are enough to make this approach work.

2.11 Primes

With the basic properties of $\mathbb{N}$ and its operations laid out, we can derive increasingly complicated concepts and prove interesting results about those concepts. We illustrate with the notion of prime numbers.

To define prime numbers precisely we begin with the concept of *divisibility*. Let **a** and **b** be natural numbers. We say that **a** *divides* **b** if there is a natural number **c** so that **ac** = **b**. The notation is **a|b**.

For example, **2|10**. Why? To show that **a** = **2** divides **b** = **10** we need to present a number **c** so that **2** · **c** = **10**; clearly that number is **5**.

On the other hand **2|9** is false because there is no natural number **c** so that **2c** = **9**.

We may also express the statement **a|b** by saying that **a** is a *divisor* of **b** or, equivalently, that **a** is a *factor* of **b**.

Here is another way to express divisibility. Suppose **a|b** and **a** ≠ **0**. Then, by definition, there is a natural number **c** so that **ac** = **b**. We can write this as **b** ÷ **a** = **c**. We don't fully address the issue of division until Chapter 5 but we can make some preliminary remarks here.

Within the restricted context of the natural numbers, sometimes **b** ÷ **a** is a natural number, but sometimes there is no answer. We know that **10** ÷ **2** = **5** but there is no natural number that we can assign to the expression **9** ÷ **2**.

In the case of **9** ÷ **2** the best we can do is give a *quotient* and a *remainder*. We know that **9** = **4** · **2** + **1**; the number **4** is the quotient and the number **1** is the remainder.

The relationship between the numbers **a** and **b** with the quotient and remainder when dividing **b** by **a** is summarized in this result.[3]

---

[3] Were the focus of this book on number theory, we would present a proof of this result here

> **Proposition 2.8** (Division Algorithm). *Let* $a, b \in \mathbb{N}$ *with* $a \neq 0$. *Then there are natural numbers* q *and* r *with* $0 \leq r < a$ *so that* $b = qa + r$.
> *Furthermore, given such* a *and* b, *the numbers* q *and* r *are unique.*

For example, suppose $a = 7$ and $b = 25$. We note that 7 goes into 25 three times and leaves a remainder of 4:

$$25 = 3 \cdot 7 + 4.$$

Here $q = 3$ and $r = 4$. The second part of Proposition 2.8 ensures us that the only numbers q and r for which $25 = q \cdot 7 + r$ and $0 \leq r < 7$ are $q = 3$ and $r = 4$.

We are ready to define prime numbers.

> **Definition 2.9.** Let $p \in \mathbb{N}$. We say p is a *prime* provided $p > 1$ and the only divisors of p are 1 and p.

Natural numbers that are greater than 1 and are not prime are called *composite*.

In this section we present two classic, important results about prime numbers. First, that every positive natural number has a unique factorization into primes. Second, that there are infinitely many primes.

The Fundamental Theorem of Arithmetic

One of the cornerstones of number theory is this classic result of Euclid.

> **Theorem 2.10** (Fundamental Theorem of Arithmetic). *Every positive natural number equals the product of prime numbers, and this factorization is unique up to the order of the primes.*

For example, consider the number 315. We note that $315 = 3 \cdot 3 \cdot 5 \cdot 7$ is a factorization of 315 into primes. We might also write that $315 = 5 \cdot 3 \cdot 7 \cdot 3$, but the only change is the order of the primes.

---

using the basic properties presented in the previous section. Instead, we simply use this result as needed.

Do prime numbers themselves factor into primes? Yes, because we allow a product with only one factor: **7** = **7**.

Ah, but what about **1**? Does **1** factor into primes? Here we have to be even more permissive with how we understand factoring. The number **1** is equal to an empty product; that is, a multiplication expression with no terms! By default, when an operation, say ⋆, has an identity element, then a ⋆-expression with no terms is defined to be the identity element for ⋆. Hence, an empty product equals **1** because **1** is the identity element for multiplication.

We prove the first part of Theorem 2.10. The uniqueness claim requires a lot more work; it is an interesting and worthwhile diversion, but it does not play a role in our story leading to the real numbers.[4]

We show that every positive natural number factors into primes. Here's the gist of the idea.

Let **n** be a natural number. If **n** = **1** then **n** factors into an empty product. If **n** itself is prime, then it is its own factorization. Otherwise, if **n** is composite, then **n** has a divisor **d** with **1** < **d** < **n**. In other words **d** and **n/d** are both natural numbers. We now repeat this same process for both **d** and **n/d** until all the factors are primes. Intuitively, we know that this process bottoms out with all prime factors. Here's how we explain that rigorously.

We want to show that all positive natural numbers can be factored into primes. To do this, we show that there are no "bad" positive natural numbers, by which we mean positive natural numbers that can't be factored into primes. In notation, let X be the set

$$X = \{\mathbf{n} \in \mathbb{N} : \mathbf{n} > 0 \text{ and } \mathbf{n} \text{ does not factor into primes}\}.$$

Our proof is complete if we show that $X = \emptyset$.

How do we show that $X = \emptyset$? The key idea is *proof by contradiction*: We imagine the scenario in which $X \neq \emptyset$ and then argue that this leads to an impossible conclusion. And that implies that the supposition $X \neq \emptyset$ is impossible.

Following this plan: Suppose $X \neq \emptyset$. By the well-ordering property (WO) X has a least element; let's call it **n**. That is, **n** is the smallest positive natural number that cannot be factored into primes. We know that **n** ≠ **1** (because **1** factors into an empty product) and **n** is not prime (because a prime is its own factorization). Hence **n** is composite. That means it has a divisor **d** with **1** < **d** < **n**. Notice that both **d** and **n/d** are positive natural numbers and both are smaller than **n**. Thus neither **d** nor **n/d** is in X, and therefore we know **d** factors into primes and **n/d** also factors into primes.

Finally, since **n** = **d** · (**n/d**) we simply combine the factorization of **d** and the factorization of **n/d** to give a factorization of **n**.

---

[4]See a book on number theory or [10].

Ah! But we said that **n** cannot be factored into primes (because **n** ∈ X) and we also know that **n** can be factored into primes (we just proved it). That's impossible!

The assumption that $X \neq \emptyset$ leads to an impossible conclusion and therefore must be false. That means $X = \emptyset$, that is, there are no exceptions to the statement *every positive integer factors into primes.*

How Many Primes?

The numbers **2**, **3**, **5**, **7**, **11**, and **13** are the first few primes. The number **7,919** is the one-thousandth prime. The number **15,485,863** is the one-millionth prime.

Clearly there are a lot of prime numbers. One may wonder, just how many primes are there? Do they go on forever? This question was answered in antiquity by none other than Euclid.

Theorem 2.11. *There are infinitely many primes.*

The key idea in the proof is to show that it is impossible for there to be only finitely many primes. This is another example of *proof by contradiction*.

We consider the hypothetical: Suppose there are only finitely many primes. What would that imply? We show that such an assumption implies something impossible: that a certain number both is and is not a prime!

Proof of Theorem 2.11. Suppose there were only finitely many primes. That would imply that there is a last prime **q** and all natural numbers greater than **q** are not prime. Consider this number:

$$\mathbf{N} = [2 \cdot 3 \cdot 5 \cdot \dots \cdot q] + 1. \qquad (*)$$

In words: **N** is the value we get when we multiply all the prime numbers together and add **1**. Since **N** is greater than any prime, it must not be a prime.

Since **N** is not a prime and $\mathbf{N} > 1$, we know that **N** is composite. We proved that every composite number factors into primes; therefore **N** is divisible by a prime.

Now we look carefully at how we defined **N** in (∗).

To begin, notice that **N** is not divisible by **2** because $\mathbf{N} = 2 \cdot$ something $+ 1$. That means if we divide **N** by **2** we would get a remainder of **1**.

Likewise, **N** is not divisible by **3** because $\mathbf{N} = 3 \cdot$ something $+ 1$. And **N** is not divisible by **5** because $\mathbf{N} = 5 \cdot$ something $+ 1$. Indeed, **N** is not divisible by any prime **p** because $\mathbf{N} = p \cdot$ something $+ 1$.

Therefore **N** is not divisible by any prime. However, we also noted that **N** is not itself a prime (it is bigger than all primes) and therefore must be divisible by a prime.

We created **N** under the assumption that there are only finitely many primes and from there we showed that **N** must be divisible by a prime and must also not be divisible by any prime. That's impossible. Hence our assumption that there only finitely many primes must be false.

Conclusion: There are infinitely many primes. □

Recap

This chapter defined natural numbers in terms of finite sets of basic objects. Specifically, natural numbers are equivalence classes of finite sets stemming from the has-a-one-to-one-correspondence-with relation.

We showed how to define addition, multiplication, and less-than-or-equal for natural numbers as well as reviewing their properties.

We offered the Peano axioms as an alternative way to define natural numbers purely in terms of their properties.

We closed with a discussion of prime numbers with an emphasis on the fact that there are infinitely many primes and that positive natural numbers have a unique factorization into primes.

Exercises

2.1 The natural number **3** is the equivalence class of some sets. Give two different representatives of **3** that are not disjoint, and two different representatives of **3** that are disjoint.

2.2 True or false: $[\![\emptyset]\!] = \{\emptyset\}$. Explain your answer.

2.3 The union and intersection of finite sets are also finite. Can the same be said of infinite sets? That is, suppose A and B are infinite sets. Is it necessarily the case that $A \cup B$ is infinite? Is it necessarily the case that $A \cap B$ is infinite?

2.4 Let **a** be a natural number and let A be a set. True or false: To assert that A is representative of **a** is the same as asserting that $A \in \mathbf{a}$.

2.5 Let $\mathbf{a}, \mathbf{b} \in \mathbb{N}$. Suppose A is a representative of **a** and B is a representative of **b**, and that $A \cap B = \emptyset$.

    a) What natural number is given by $[\![A \cup B]\!]$?

    b) What natural number is given by $[\![A \cap B]\!]$?

    c) What natural number is given by $[\![A - B]\!]$?

    d) What natural number is given by $[\![A \times B]\!]$?

2.6 Let a, b, c, d, and e be distinct objects. Find a one-to-one correspondence between the sets $\{a, b\} \times \{c, d, e\}$ and $\{c, d, e\} \times \{a, b\}$. Note that this implies that $\mathbf{2} \cdot \mathbf{3} = \mathbf{3} \cdot \mathbf{2}$.

2.7 Show that $\mathbf{4} + \mathbf{4} = \mathbf{2} \cdot \mathbf{4}$ by the following method:

   a) Create a representative S of $\mathbf{4} + \mathbf{4}$ from two disjoint representatives of $\mathbf{4}$.

   b) Create a representative T of $\mathbf{2} \cdot \mathbf{4}$ from a representative of $\mathbf{2}$ and a representative of $\mathbf{4}$.

   c) Construct a one-to-one correspondence between S and T.

   d) From that correspondence, conclude that $\mathbf{4} + \mathbf{4} = \mathbf{2} \cdot \mathbf{4}$.

2.8 Let $\mathbf{a} \in \mathbb{N}$. Show that $\mathbf{a} + \mathbf{a} = \mathbf{2} \cdot \mathbf{a}$ using the same approach as in Exercise 2.7.

2.9 Note that $\{a\}$ is a finite set because it is a singleton. Is $[\![\{a\}]\!]$ a finite set?

2.10 Let $\mathbf{a}, \mathbf{b} \in \mathbb{N}$. Show, using Definition 2.7, that $\mathbf{a} \leq \mathbf{a} + \mathbf{b}$.

2.11 Let $\mathbf{a}, \mathbf{b} \in \mathbb{N}$ with $\mathbf{a} \leq \mathbf{b}$. Use Definition 2.7 to show that there are representatives A of $\mathbf{a}$ and B of $\mathbf{b}$ with $A \subseteq B$.

2.12 Let $A = \{a, b\}$, $B = \{c, d, e\}$, and $C = \{x, y\}$.

   a) Write out the set $A \times B$ and $A \times C$.

   b) Using that result, write out the full set $(A \times B) \cup (A \times C)$.

   c) Write out the set $B \cup C$.

   d) Using that result, write out the set $A \times (B \cup C)$.

   e) Check that $(A \times B) \cup (A \times C) = A \times (B \cup C)$.

2.13 Let $\mathbf{a}, \mathbf{b} \in \mathbb{N}$ with $\mathbf{a} \leq \mathbf{b}$. Create a definition of $\mathbf{b} - \mathbf{a}$ using representatives of $\mathbf{a}$ and $\mathbf{b}$.

2.14 Let A and B be sets of basic objects. Is $A \times B$ also a set of basic objects?

In the next two exercises we take advantage of the fact that natural numbers are the answers to counting problems. Multiplication $\mathbf{a} \cdot \mathbf{b}$ corresponds to counting the elements of the Cartesian product $A \times B$ where $\mathbf{a}$ is the number of elements in set A and $\mathbf{b}$ is the number of elements in B.

A great way to visualize this is by drawing a checkerboard with $\mathbf{a}$ rows and $\mathbf{b}$ columns. The number of squares in the checkerboard is $\mathbf{a} \cdot \mathbf{b}$.

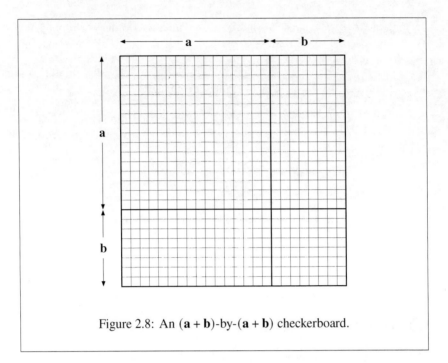

Figure 2.8: An $(a+b)$-by-$(a+b)$ checkerboard.

2.15 The number of squares in a checkerboard that has $a + b$ squares on each side is $(a+b) \cdot (a+b) = (a+b)^2$. See Figure 2.8.

Note that the checkerboard is subdivided into four regions. Adding the number of squares in each of the four regions, derive an alternative expression for $(a+b)^2$.

2.16 Suppose $a \geq b$ and consider the a-by-a checkerboard depicted in Figure 2.9. Use counting to justify the algebraic inequality $a^2 + b^2 \geq 2ab$.

2.17 Explain how to define the relations $\leq$, $>$, and $\geq$ based on the relation $<$.

2.18 Using only properties (L1)–(L4) (see page 38) demonstrate the *trichotomy* property: If a and b are natural numbers then exactly one of the following is true: $a < b$, $a = b$, or $a > b$. See the box on page on page 38 for how $<$ and $>$ are defined from $\leq$.

2.19 For which natural numbers a do we have $a|0$? For which $a \in \mathbb{N}$ do we have $0|a$?

2.20 The is-a-factor-of relation on $\mathbb{N}$ is denoted by the symbol $|$. Show that $|$ is reflexive, antisymmetric, and transitive.

2.21 For each of the following sets, determine the least element.

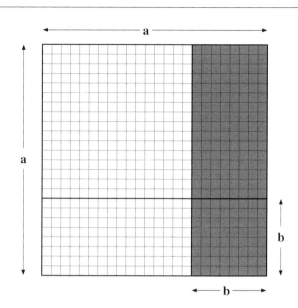

Figure 2.9: An **a**-by-**a** checkerboard that contains a **b**-by-**b** checkerboard on the lower right. The filled squares show an **a**-by-**b** rectangle contained in the checkerboard that overlaps the **b**-by-**b** square.

   a) Prime numbers.
   b) Composite numbers.
   c) Positive natural numbers that are neither prime nor composite.
   d) Natural numbers whose base-ten representation has two digits.

2.22 Use the well-ordering property to explain why there cannot be an infinitely long decreasing list of natural numbers: $a_1 > a_2 > a_3 > \cdots$.

2.23 For each of the following subsets of $\mathbb{N}$ determine the sum of the elements in the set as well as the product of the elements of the set.

   a) $\{2, 4, 6\}$.
   b) $\{x \in \mathbb{N} : x \text{ is prime and } x < 10\}$.
   c) Even prime numbers.
   d) ∅.

2.24 A natural number **n** is called *even* provided **n = 2a** for some natural number **a**. (This is a restricted version of Definition 0.1.)

Use Proposition 2.8, the Division Algorithm, to show that if a natural number **n** is not even, then **n = 2a + 1** for some natural number **a**.

In other words, if a natural number is not even, then it must be odd (using the definition of *odd* presented in Exercise 0.7).

2.25 The Peano axioms [given $\mathbb{N}$, next, and properties (P1)–(P3)] assume the existence of the number **0**. How would the number **1** be defined in this system?

2.26 Show how to define multiplication using the Peano axioms.

2.27 The common place-value notation for natural numbers consists of a (finite) sequence of one or more of the digits 0, 1, 2, 3, 4, 5, 6, 7, 8, and 9. Unless the sequence is only a single digit, the first element in the sequence may not be a 0. For example, 3,506 means

$$3 \cdot 10^3 + 5 \cdot 10^2 + 0 \cdot 10^1 + 6 \cdot 10^0.$$

This is known as *base-ten* notation.

Aside from familiarity, there's nothing particularly special about base ten, and we can readily express natural numbers in another base. For example, in base five we use the digits 0 through 4 with various powers of 5. For example, $3,204_{\text{FIVE}}$ means

$$3 \cdot 5^3 + 2 \cdot 5^2 + 0 \cdot 5^1 + 4 \cdot 5^4.$$

This equals 429 in base ten. To be especially clear about the base under consideration, we can append the subscript TEN and write $3,204_{\text{FIVE}} = 429_{\text{TEN}}$.

    a) Express 123_{FIVE} and 123_{SEVEN} in base ten.

    b) Express the natural number 145_{TEN} in bases five and seven.

    c) Calculate $62_{\text{SEVEN}} + 33_{\text{SEVEN}}$ and express your answer in base seven. Work directly in base seven. That is, do not convert to base ten, add, and then convert back to base seven.

    d) Calculate $23_{\text{FIVE}} \cdot 24_{\text{FIVE}}$ and express your answer in base five. Work directly in base five.

2.28 Is $\{1\} \in 1$?

2.29 Definition 2.1 creates the world of basic objects from two primordial basic objects: *a* and *b*. Is this necessary? That is, might we have a definition of basic object that begins with only a single basic object, *a*?

2.30 Let A and B be sets all of whose elements are basic objects. Are the elements of $A \times B$ also basic objects?

Chapter 3

$\mathbb{Z}$: Integers

There are two intermediate stops on our journey from the natural numbers to the real numbers: the integers and the rational numbers. This chapter builds on the natural numbers, $\mathbb{N}$, to construct the integers.

The transitions through integers and rational numbers can be thought of as providing repairs to two "defects" that plague the natural numbers: we can't subtract and we can't divide. Of course, sometimes we can subtract (e.g., $12 - 4 = 8$) and sometimes we can divide (e.g., $12 \div 4 = 3$), but neither $4 - 12$ nor $4 \div 12$ are meaningful in the limited context of the natural numbers.

The motivation to extend the natural numbers to integers can be expressed in terms of an equation to solve. Given $a, b \in \mathbb{N}$ we want to solve the simple equation $a+x = b$. Sometimes there is a solution to this equation (e.g., $4+x = 12$) and sometimes there isn't (e.g., $12 + x = 4$).

3.1 Reintroducing the Integers

Thanks to our work in the previous chapters, we have, in hand, the natural numbers $\mathbb{N} = \{0, 1, 2, 3, \ldots\}$. A familiar way to define the integers is simply to extend $\mathbb{N}$ with negative numbers. That is, for every positive value n in $\mathbb{N}$, append a new number named $-n$. The set of integers, denoted by $\mathbb{Z}$, is then simply this:

$$\mathbb{Z} = \{0, 1, 2, 3, \ldots\} \cup \{-1, -2, -3, \ldots\} = \{\ldots, -3, -2, -1, 0, 1, 2, 3, \ldots\}.$$

The next step would be to define how addition, multiplication, and less-than-or-equal need to be extended to be consistent with all the usual algebraic properties.

On the one hand, this would be familiar and appear to be simple. On the other hand, it's not so simple. Why do we require the product of two negative numbers to be positive, but the product of a negative and a positive to be negative? (See Exercise 3.15.) We need to describe how 0 (which is neither positive nor negative) behaves with the newly minted negative values. The reason for all this fuss is that we want the familiar algebraic properties for natural numbers to apply to all integers. This approach takes a good deal of effort to secure all the details.

We take another approach. At first glance, it may seem that our approach is too complicated. However, it is reminiscent of how we transitioned from finite sets to natural numbers, and we employ this method again in the next chapter to develop rational numbers from integers. It also makes the verification of the various algebraic properties easier as it does not require us to treat positive, negative, and zero values separately.

Here is a bird's eye view of the approach. Recall from the previous chapter that we used two ingredients to construct natural numbers: finite sets and the has-a-one-to-one-correspondence-with relation. Natural numbers are the equivalence classes of this relation. We use the same recipe here, but with different ingredients.

3.2 An Equivalence Relation for Pairs of Natural Numbers

In this chapter the symbol $\equiv$ is used for an entirely new relation that applies to ordered pairs of natural numbers: (a, b) with $a, b \in \mathbb{N}$. In other words, $\equiv$ is a relation defined from $\mathbb{N} \times \mathbb{N}$ to itself.

> **Definition 3.1.** For $a, b, c, d \in \mathbb{N}$ define $(a, b) \equiv (c, d)$ provided $a + d = b + c$.

For example $(10, 3) \equiv (12, 5)$ because $10 + 5 = 3 + 12$. However $(5, 8) \not\equiv (12, 3)$ because $5 + 3 \neq 8 + 12$.

Although we do not have a definition for subtraction (yet) we informally note that $(a, b) \equiv (c, d)$ if the difference between first and second entries in the pairs is the same because $a + d = b + c$ is equivalent to $a - b = c - d$.

Here is another way to express this without mention of subtraction.

> **Proposition 3.2.** For natural numbers a, b, c, d if $(a, b) \equiv (c, d)$ then there is a natural number t such that either $(a, b) = (c + t, d + t)$ or $(c, d) = (a + t, b + t)$.

Proof. We have either $a \leq c$ or $a \geq c$.

In the first case, with $a \leq c$, we have $c = a + t$ for some $t \in \mathbb{N}$. We rewrite $(a, b) \equiv (c, d)$ as $(a, b) \equiv (a + t, d)$. This gives

$$a + d = b + a + t,$$

which gives $d = b + t$ by the cancellation property (A5). In other words, $(c, d) = (a + t, b + t)$.

When $a \geq c$ a similar argument shows that $(a, b) = (c + t, d + t)$ for some $t \in \mathbb{N}$. □

Next we check that $\equiv$ is an equivalence relation on $\mathbb{N} \times \mathbb{N}$:

- $\equiv$ is reflexive: Let $(a, b) \in \mathbb{N} \times \mathbb{N}$. Then $(a, b) \equiv (a, b)$ because $a + b = b + a$.

  [Example: $(5, 9) \equiv (5, 9)$ because $5 + 9 = 9 + 5$.]

- $\equiv$ is symmetric: Let $(a, b), (c, d) \in \mathbb{N} \times \mathbb{N}$ and suppose we know that $(a, b) \equiv (c, d)$. That means that $a + d = b + c$, which can be rewritten as $c + b = d + a$, which gives $(c, d) \equiv (a, b)$.

  [Example: $(9, 3) \equiv (12, 6)$ because $9 + 6 = 3 + 12$. However, it is also the case that $12 + 3 = 6 + 9$, which implies $(12, 6) \equiv (9, 3)$.]

  Please note that the symmetric property does *not* say that $(a, b) \equiv (b, a)$. Indeed, $(2, 5) \not\equiv (5, 2)$ because $2 + 2 \neq 5 + 5$.

- $\equiv$ is transitive: Let $(a, b), (c, d), (e, f) \in \mathbb{N} \times \mathbb{N}$ with $(a, b) \equiv (c, d)$ and $(c, d) \equiv (e, f)$. These give the following equations:

  $$(a, b) \equiv (c, d) \implies a + d = b + c,$$
  $$(c, d) \equiv (e, f) \implies c + f = d + e.$$

  Adding these two equations gives:

  $$a + d + c + f = b + c + d + e$$
  $$\implies (a + f) + (c + d) = (b + e) + (c + d).$$

  The second equation is of the form $(a + f) + X = (b + e) + X$. By the cancellation property (A5) it follows that $a + f = b + e$ and so $(a, b) \equiv (e, f)$.

  [Example: $(10, 2) \equiv (8, 0)$ because $10 + 0 = 2 + 8$ and $(8, 0) \equiv (13, 5)$ because $8 + 5 = 0 + 13$. Notice now that we also have $(10, 2) \equiv (13, 5)$ because $10 + 5 = 2 + 13$.]

We have established that $\equiv$ is an equivalence relation. Therefore the $\equiv$-equivalence classes form a partition of $\mathbb{N} \times \mathbb{N}$. We have a name for those equivalence classes: We call them *integers*!

3.3 Integers and their Operations

In Chapter 2 we presented the natural numbers as equivalence classes of finite sets. Following that motif, *integers* are also constructed as equivalence classes, but in this case the objects are pairs of integers and the relation $\equiv$ is the one defined in Definition 3.1.

Definition 3.3. An *integer* is an $\equiv$-equivalence class.

For example, $[\![(7, 11)]\!]$ is an integer. Let's look at this integer under a microscope. The notation $[\![(7, 11)]\!]$ stands for the set of all pairs of natural numbers that are equivalent to $(7, 11)$. For example, $(8, 12) \equiv (7, 11)$ so $(8, 12)$ is an element of $[\![(7, 11)]\!]$. Likewise, $(10, 14) \equiv (7, 11)$ and so $(10, 14) \in [\![(7, 11)]\!]$. Let's write out the set $[\![(7, 11)]\!]$ in full (or, as fully as we can since the set is infinite):

$$\{(0, 4), (1, 5), (2, 6), (3, 7), (4, 8), (5, 9), (6, 10), \underline{(7, 11)}, (8, 12), (9, 13), \ldots\}.$$

Here are some additional examples of integers: $[\![(5, 2)]\!]$, $[\![(3, 9)]\!]$, $[\![(5, 5)]\!]$, and $[\![(5, 11)]\!]$. Incidentally, these are only three different integers! Can you spot the repeat?

Of course, these look *nothing* like "ordinary" integers, but let's keep going.

Addition

We begin by defining integer addition.

Definition 3.4 (Addition of integers). Let $\mathbf{x} = [\![(a, b)]\!]$ and $\mathbf{y} = [\![(c, d)]\!]$ be integers. Their sum is defined as

$$\mathbf{x} + \mathbf{y} = [\![(a + c, b + d)]\!].$$

Note: Integers, because they are new in this chapter, are written in **boldface** and natural numbers are rendered with standard weight characters; see the boxed comment on page 28.

We need to check that Definition 3.4 is consistent because an integer $\mathbf{x}$ may represented by different ordered pairs; we need to be sure that the choice of ordered pair does not matter.

> **What's Up with this Crazy Definition of Integers?**
>
> Definition 3.3 defines an *integer* as an equivalence class of ordered pairs of natural numbers. This seems horribly complicated. What's really going on?
>
> We don't have negative numbers yet; this approach allows us to improvise. The pair (a, b) encodes the integer $a - b$. When $a \geq b$ this is a natural number, but when $a < b$ the result is negative. To represent the integer -4 we have a lot of choices. We can see -4 as $0-4$, in which case the encoding is $(0, 4)$. But -4 also equals $8 - 12$, so $(8, 12)$ is another encoding. And $(1, 5)$ is yet another encoding. Which do we use for -4? All of them! That's because $(0, 4) \equiv (8, 12) \equiv (1, 5)$. The integer -4 is the set of all pairs (a, b) with a being 4 less than b; in symbols: $-4 = [\![(0, 4)]\!]$.
>
> When do (a, b) and (c, d) encode the same integer? That happens when $a - b = c - d$. But this subtraction is not meaningful prior to defining integers. So we replace the equation $a - b = c - d$ with the equivalent $a + d = b + c$, which is exactly the condition for $(a, b) \equiv (c, d)$ in Definition 3.1.

Specifically, suppose we have different representatives for **x** and different representatives for **y**:

$$\mathbf{x} = [\![(a, b)]\!] = [\![(a', b')]\!] \quad \text{and} \quad \mathbf{y} = [\![(c, d)]\!] = [\![(c', d')]\!].$$

We need to check that $(a + c, b + d) \equiv (a' + c', b' + d')$: that is, $(a + c, b + d)$ and $(a' + c', b' + d')$ are representatives of the same integer.

Here is what we are given:

$$(a, b) \equiv (a', b') \implies a + b' = b + a',$$
$$(c, d) \equiv (c', d') \implies c + d' = d + c'.$$

Adding the two equations on the right gives

$$a + b' + c + d' = b + a' + d + c'$$
$$\implies (a + c) + (b' + d') = (a' + c') + (b + d)$$
$$\implies (a + c, b + d) \equiv (a' + c', b' + d')$$

as required.

What we have shown is that when we add **x** and **y** it doesn't matter which representatives we pick from the classes; the result will be the same. In other words, Definition 3.4 properly defines an operation for integers (which are equivalence classes of $\mathbb{N} \times \mathbb{N}$).

Multiplication

The next step is to define the product of two integers, $\mathbf{x} = [\![(a,b)]\!]$ and $\mathbf{y} = [\![(c,d)]\!]$. Given the "guilty knowledge" that $[\![(a,b)]\!]$ is a stand-in for $a - b$ and $[\![(c,d)]\!]$ is a stand-in for $c - d$ (see the boxed comment on the previous page) the following definition is not too surprising.

Definition 3.5 (Multiplication of integers). Let $\mathbf{x} = [\![(a,b)]\!]$ and $\mathbf{y} = [\![(c,d)]\!]$ be integers. Their product $\mathbf{x} \cdot \mathbf{y}$ is defined to be

$$\mathbf{x} \cdot \mathbf{y} = [\![(ac + bd, ad + bc)]\!].$$

As before, we need to check that if we pick different representatives for $\mathbf{x}$ and for $\mathbf{y}$ we arrive at the same result. This requires a lot more algebra, but the idea is the same.

It is handy to have, as a temporary notation, an operation $\otimes$ for pairs of natural numbers that underlies Definition 3.5. Let

$$(a,b) \otimes (c,d) = (ac + bd, ad + bc).$$

Definition 3.5 can then be written like this:

$$\mathbf{x} \cdot \mathbf{y} = [\![(a,b)]\!] \cdot [\![(c,d)]\!] = [\![(a,b) \otimes (c,d)]\!].$$

Note that $\otimes$ is a commutative and associative operation; see Exercise 3.10.

We need to show that the result of multiplying two integers does not depend on the representatives chosen. To that end, we start with a special case.

Proposition 3.6. Let a, b, c, d, c', d' be natural numbers with $(c,d) \equiv (c', d')$. Then $(a,b) \otimes (c,d) \equiv (a,b) \otimes (c', d')$.

Proof. To see that this is correct we use Proposition 3.2. We replace (c', d') with $(c+t, d+t)$ for some natural number t. (Note: This works if $c \leq c'$. If $c > c'$ simply reverse the roles of (c,d) and (c', d').)

We now calculate:

$$(a,b) \otimes (c,d) = (ac + bd, ad + bc),$$

$$\begin{aligned}
(a,b) \otimes (c',d') &= (a,b) \otimes (c+t, d+t) \\
&= \Big(a(c+t) + b(d+t), a(d+t) + b(c+t)\Big) \\
&= (ac + at + bd + bt, ad + at + bc + bt) \\
&= \Big(ac + bd + (at+bt), ad + bc + (at+bt)\Big) \\
&= (ac + bd + T, ad + bc + T) \quad \text{where } T = at + bt \\
&\equiv (ac + bd, ad + bc) \\
&= (a,b) \otimes (c,d). \quad \square
\end{aligned}$$

Now we argue that when multiplying integers **x** and **y** the result does not depend on which representatives we choose. As before, suppose we have

$$\mathbf{x} = [\![(a,b)]\!] = [\![(a',b')]\!] \quad \text{and} \quad \mathbf{y} = [\![(c,d)]\!] = [\![(c',d')]\!].$$

To that end, we show that $[\![(a,b) \otimes (c,d)]\!]$ equals $[\![(a',b') \otimes (c',d')]\!]$.

By Proposition 3.6 we know

$$(a,b) \otimes (c,d) \equiv (a,b) \otimes (c',d') \qquad (*)$$

Using the fact that $\otimes$ is commutative and applying Proposition 3.6 again we have

$$(a,b) \otimes (c',d') = (c',d') \otimes (a,b) \equiv (c',d') \otimes (a',b'). \qquad (**)$$

Combining equations $(*)$ and $(**)$ we see that $(a,b) \otimes (c,d) \equiv (a',b') \otimes (c',d')$ from which it follows that $[\![(a,b) \otimes (c,d)]\!] = [\![(a',b') \otimes (c',d')]\!]$. Therefore multiplication is well defined.

Less-than-or-equal

Just as we defined addition and multiplication for natural numbers, we must also define the $\leq$ relation for integers.

Definition 3.7 ($\leq$ for integers). Let **x** and **y** be integers with $\mathbf{x} = [\![(a,b)]\!]$ and $\mathbf{y} = [\![(c,d)]\!]$. We say that **x** *is less than or equal to* **y**, and we write $\mathbf{x} \leq \mathbf{y}$ provided $a + d \leq b + c$.

Please note that we have used the same symbol $\leq$ twice in the definition. In the first case ($\mathbf{x} \leq \mathbf{y}$), $\leq$ is a new relation we are defining for integers. In the second case ($a+d \leq b+c$), the symbol stands for the less-than-or-equal relation for natural numbers, which is already defined.

As in the case of addition and multiplication, this definition does not depend on which representatives for **x** and **y** we choose. See Exercise 3.11.

We also have the trichotomy property for integers.

Proposition 3.8. *Let* **x** *and* **y** *be integers. Then exactly one of the following is true:* **x** < **y**, **x** = **y**, *or* **x** > **y**.

To see why this works, let $\mathbf{x} = [\![(a,b)]\!]$ and $\mathbf{y} = [\![(c,d)]\!]$. We then compare $a+d$ and $b+c$:

- If $a+d < b+c$ then **x** < **y**.

- If $a+d = b+c$ then **x** = **y**.

- If $a+d > b+c$ then **x** > **y**.

In this way, the trichotomy property for integers derives easily from the trichotomy property for natural numbers.

This also gives us a simple way to distinguish positive and negative integers. As usual, an integer **x** is *positive* if **x** > **0** and is *negative* if **x** < **0**.

Let (a,b) represent **x**. When **x** is positive we have $a > b$. Since $a > b$ we can write $a = b + t$ where t is a nonzero natural number. Notice that $(a,b) = (b+t, b) \equiv (t, 0)$ because $(b+t) + 0 = b+t$.

On the other hand, if **x** < **0** we have $a < b$ and so we can write $b = a + s$ where s is a nonzero natural number. Then $(a,b) = (a, a+s) \equiv (0, s)$.

Finally, we know that $\mathbf{0} = [\![(0,0)]\!]$. Collecting these thoughts together, we have the following:

Proposition 3.9. *Let* **x** *be an integer.*

- *If* **x** > **0** *then* $\mathbf{x} = [\![(a, 0)]\!]$ *for some nonzero natural number a.*

- *If* **x** = **0** *then* $\mathbf{x} = [\![(0, 0)]\!]$.

- *If* **x** < **0** *then* $\mathbf{x} = [\![(0, a)]\!]$ *for some nonzero natural number a.*

3.4 Arithmetic Properties of the Integers

The full set of integers is denoted by the symbol $\mathbb{Z}$. The algebraic properties of integer addition and multiplication are undoubtedly familiar to you. We present them here for reference and then discussion.

Proposition 3.10. *The integers $\mathbb{Z}$, together with the operations of addition and multiplication, have the following properties:*

- *Addition is commutative: For all $\mathbf{x}, \mathbf{y} \in \mathbb{Z}$ we have $\mathbf{x} + \mathbf{y} = \mathbf{y} + \mathbf{x}$.*

- *Addition is associative: For all $\mathbf{x}, \mathbf{y}, \mathbf{z} \in \mathbb{Z}$ we have $(\mathbf{x} + \mathbf{y}) + \mathbf{z} = \mathbf{x} + (\mathbf{y} + \mathbf{z})$.*

- *Identity element for addition: There is an element $\mathbf{0} \in \mathbb{Z}$ with the property that for all $\mathbf{x} \in \mathbb{Z}$ we have $\mathbf{x} + \mathbf{0} = \mathbf{x}$.*

- *Additive inverses: For every $\mathbf{x} \in \mathbb{Z}$ there is an element $-\mathbf{x} \in \mathbb{Z}$ with the property $-\mathbf{x} + \mathbf{x} = \mathbf{0}$.*

- *Multiplication is commutative: For all $\mathbf{x}, \mathbf{y} \in \mathbb{Z}$ we have $\mathbf{x} \cdot \mathbf{y} = \mathbf{y} \cdot \mathbf{x}$.*

- *Multiplication is associative: For all $\mathbf{x}, \mathbf{y}, \mathbf{z} \in \mathbb{Z}$ we have $(\mathbf{x} \cdot \mathbf{y}) \cdot \mathbf{z} = \mathbf{x} \cdot (\mathbf{y} \cdot \mathbf{z})$.*

- *Identity element for multiplication:s There is an element $\mathbf{1} \in \mathbb{Z}$, $\mathbf{0} \neq \mathbf{1}$, with the property that for all $\mathbf{x} \in \mathbb{Z}$ we have $\mathbf{1} \cdot \mathbf{x} = \mathbf{x}$.*

- *Distributive property: For all $\mathbf{x}, \mathbf{y}, \mathbf{z} \in \mathbb{Z}$ we have $\mathbf{x} \cdot (\mathbf{y}+\mathbf{z}) = \mathbf{x} \cdot \mathbf{y} + \mathbf{x} \cdot \mathbf{z}$.*

The mathematical terminology that encompasses all of the properties in Proposition 3.10 is that $(\mathbb{Z}, +, \cdot)$ is a *commutative ring*.

That addition and multiplication are both commutative and associative is the subject of Exercises 3.6–10. Let's see if we can find integer representations that satisfy the properties for numbers $\mathbf{0}$ and $\mathbf{1}$.

What natural numbers a and b give us an integer $[\![(a, b)]\!]$ that behaves like zero? A good (and correct) guess is $(0, 0)$. Let's check that this is correct.

Let $\mathbf{0} = [\![(0,0)]\!]$ and let $\mathbf{x} = [\![(c,d)]\!]$ be an integer. What is $\mathbf{0} + \mathbf{x}$? It is:

$$\mathbf{0} + \mathbf{x} = [\![(0,0)]\!] + [\![(c,d)]\!] = [\![(0+c, 0+d)]\!] = [\![(c,d)]\!] = \mathbf{x}$$

as required.

Incidentally, written in full $\mathbf{0}$ is this:

$$\mathbf{0} = [\![(0,0)]\!] = \{(0,0), (1,1), (2,2), (3,3), \ldots\} = \{(n,n) : n \in \mathbb{N}\}.$$

A familiar property of zero is that multiplying by zero always gives zero. Let's see if that works here. If $\mathbf{x} = [\![(a,b)]\!]$, then

$$\mathbf{0} \cdot \mathbf{x} = [\![(0,0)]\!] \cdot [\![(a,b)]\!] = [\![(0 \cdot a + 0 \cdot b, 0 \cdot b + 0 \cdot a)]\!] = [\![(0,0)]\!] = \mathbf{0}.$$

For a bit of added fun, we redo the same calculation using the fact that $\mathbf{0}$ also equals $[\![(1,1)]\!]$:

$$\mathbf{0} \cdot \mathbf{x} = [\![(1,1)]\!] \cdot [\![(a,b)]\!] = [\![(1 \cdot a + 1 \cdot b, 1 \cdot b + 1 \cdot a)]\!] = [\![(a+b, a+b)]\!] = \mathbf{0}.$$

Proposition 3.10 also speaks of additive inverses. What is the additive inverse of $\mathbf{x} = [\![(a,b)]\!]$? We need to find an integer $[\![(c,d)]\!]$ that, when added to $[\![(a,b)]\!]$, gives $\mathbf{0}$. That is, $\mathbf{0} = [\![(a+c, b+d)]\!]$. To make this work, we must have $a + c = b + d$, and a simple solution to that is $(c,d) = (b,a)$. Let's see if that's right:

$$[\![(a,b)]\!] + [\![(b,a)]\!] = [\![(a+b, b+a)]\!] = \mathbf{0}.$$

In other words, if $\mathbf{x} = [\![(a,b)]\!]$ then $-\mathbf{x}$ is simply $[\![(b,a)]\!]$.

In this context the minus sign refers to the additive inverse of a number. For example, $-\mathbf{3} = [\![(0,3)]\!]$ is the additive inverse of $\mathbf{3} = [\![(3,0)]\!]$. Of course, the minus sign is also used for subtraction:

Definition 3.11 (Subtraction). Let $\mathbf{x}$ and $\mathbf{y}$ be integers. Then $\mathbf{x} - \mathbf{y}$ is the integer $\mathbf{x} + (-\mathbf{y})$.

Next we need to find an identity element for multiplication; that is, we need to determine the representation of $\mathbf{1}$. A reasonable guess: $\mathbf{1} = [\![(1,0)]\!]$. Let's see if this is correct. Let $\mathbf{x} = [\![(a,b)]\!]$ and multiply:

$$\mathbf{1} \cdot \mathbf{x} = [\![(1,0)]\!] \cdot [\![(a,b)]\!] = [\![(1,0) \otimes (a,b)]\!] = [\![(1 \cdot a + 0 \cdot b, 1 \cdot b + 0 \cdot a)]\!]$$
$$= [\![(a,b)]\!] = \mathbf{x}.$$

It's easy to check that $\mathbf{0} \neq \mathbf{1}$. Finally, the distributive property holds; see Exercise 3.14.

Ordering and Arithmetic

The integers, together with the operations of addition and multiplication, form a commutative ring. They also have an ordering relation, $\leq$. This ordering interacts with the arithmetic of $+$ and $\cdot$ in the expected ways.

> **Proposition 3.12.** *Let* **x**, **y**, *and* **z** *be integers.*
>
> - *If* **x** < **y**, *then* **x** + **z** < **y** + **z**.
> - *If* **0** < **x** *and* **0** < **y**, *then* **0** < **x** · **y**.

3.5 Assimilation

The natural numbers serve as the building blocks for the integers. To keep the distinction between natural numbers and integers clear, in this chapter we use lightweight type (such as n or 5) to represent elements of $\mathbb{N}$ and boldface type (**x** or **−2**) for the elements of $\mathbb{Z}$, the integers.

However, nestled inside the integers we find a copy of the natural numbers. That is, the nonnegative integers **0**, **1**, **2**, **3**, and so on are an exact replica of their more basic cousins 0, 1, 2, 3, and so forth. More explicitly, if $\mathbf{a} = [\![(a,0)]\!]$, $\mathbf{b} = [\![(b,0)]\!]$, and $\mathbf{c} = [\![(c,0)]\!]$ then we have this:

$$\begin{aligned} \mathbf{a} + \mathbf{b} = \mathbf{c} &\leftrightarrow a + b = c, \\ \mathbf{a} \cdot \mathbf{b} = \mathbf{c} &\leftrightarrow a \cdot b = c, \\ \mathbf{a} \leq \mathbf{b} &\leftrightarrow a \leq b. \end{aligned}$$

Going forward, we have no need to use the original definition of $\mathbb{N}$. We can find a copy of $\mathbb{N}$ in the new set $\mathbb{Z}$ that we just created.

Henceforth, the natural numbers are simply part of the integers and we write $\mathbb{N} \subseteq \mathbb{Z}$.

Recap

We defined an equivalence relation $\equiv$ on pairs of natural numbers with $(a,b) \equiv (c,d)$ provided $a + d = b + c$. Integers are defined as the equivalence classes of this relation. From there, we defined addition, multiplication, and the less-than-or-equal relation for integers. Finally, we noted that a copy of the natural numbers can be found embedded in the integers and so, going forward, we need to use only integers and relegate the natural numbers as a subset.

Exercises

3.1 Two of these integers are the same: $[\![(5,2)]\!]$, $[\![(3,9)]\!]$, $[\![(5,5)]\!]$, and $[\![(5,11)]\!]$. Which are they?

3.2 The integer **5** is the set $[\![(5,0)]\!]$. This is an infinite set. Write it down in the form
$$\{(5,0), (\square,\square), (\square,\square), (\square,\square), \ldots\}$$
including at least three ordered pairs after $(5,0)$.

Write -5 in the same way.

3.3 Choose two different representatives a and b of **5** [other than $(5,0)$] and two different representatives c and d of -3 [other than $(0,3)$] and calculate $5 + (-3)$ four different ways using $a+c$, $a+d$, $b+c$, and $b+d$. Show that you get the same answer in all cases. This illustrates that the result of $5 + -3$ does not depend on the choice of representatives.

Repeat this with multiplication instead of addition to illustrate that the result of $(5) \cdot (-3)$ does not depend on the choice of representatives.

3.4 What is the relationship between the integers $[\![(a,b)]\!]$ and $[\![(b,a)]\!]$?

3.5 Let $a, b \in \mathbb{N}$. Calculate $(a,b) \otimes (a,b)$ and $(a,b) \otimes (b,a)$, and compare the results. What does this illustrate?

3.6 Show that integer addition is commutative and associative.

3.7 The integer **1** is the set $[\![(1,0)]\!] = \{(1,0), (2,1), (3,2), \ldots\}$, which can be written using set builder notation as
$$\mathbf{1} = \{(a+1, a): a \in \mathbb{N}\}.$$

a) Write the integer **2** using set builder notation.

b) Show that $1 + 1 = 2$ by choosing any two representatives of **1**, $(a+1, a)$ and $(b+1, b)$, and then applying Definition 3.4.

c) Similarly, show that $1 \cdot 1 = 1$ by choosing any two representatives of **1**, $(a+1, a)$ and $(b+1, b)$, and then applying Definition 3.5.

3.8 The integer **0** is $\{(n,n): n \in \mathbb{N}\}$. Let **x** be any integer. Show, by choosing a representative of **0** and a representative of **x**, that $\mathbf{0} \cdot \mathbf{x} = \mathbf{0}$.

3.9 Let **x** be any integer. Show that $2 \cdot \mathbf{x} \neq \mathbf{1}$.

3.10 Show that the $\otimes$ operation (defined on page 58) is commutative and associative. This implies that integer multiplication is commutative and associative.

3.11 Choose two different representatives of **5** [other than $(5,0)$] and two different representatives **8** [other than $(8,0)$].

Show that, no matter which representatives we select, Definition 3.7 shows that $\mathbf{5} \leq \mathbf{8}$.

3.12 Let **x** and **y** be nonzero integers. Show that $\mathbf{x} \cdot \mathbf{y} \neq \mathbf{0}$.

3.13 Verify these cancellation properties for integers **a**, **b**, and **x**:

   a) If $\mathbf{a} + \mathbf{x} = \mathbf{b} + \mathbf{x}$ then $\mathbf{a} = \mathbf{b}$.

   b) If $\mathbf{x} \neq \mathbf{0}$ and $\mathbf{a} \cdot \mathbf{x} = \mathbf{b} \cdot \mathbf{x}$ then $\mathbf{a} = \mathbf{b}$.

3.14 Verify the distributive property for integers: $\mathbf{x} \cdot (\mathbf{y} + \mathbf{z}) = \mathbf{x} \cdot \mathbf{y} + \mathbf{x} \cdot \mathbf{z}$.

3.15 Show that multiplying two positive integers or two negative integers results in a positive integer. Show that multiplying a positive and a negative integer results in a negative integer.

3.16 Show that the sum of two positive integers is positive and that the sum of two negative integers is negative.

3.17 Chapter 2 lists several properties enjoyed by the natural numbers: (A1)–(A5), (M1)–(M4), (D), (L1)–(L4), (LA), (LM), and (WO). Which of these properties hold, and which do not hold, when we replace $\mathbb{N}$ by $\mathbb{Z}$?

3.18 Polynomials are algebraic expressions formed by combining numbers and a variable x using the operations of addition and multiplication. The notation $\mathbb{Z}[x]$ stands for the set of all polynomials whose coefficients are integers, that is, all expressions of the form

$$\mathbf{a}_n x^n + \mathbf{a}_{n-1} x^{n-1} + \cdots + \mathbf{a}_1 x + \mathbf{a}_0$$

where the $\mathbf{a}_i$s are integers.

Polynomials can be added and multiplied. Do they form a commutative ring? (See Proposition 3.10.)

Which feature(s) of the integers do not apply to $\mathbb{Z}[x]$?

Chapter 4

$\mathbb{Z}_m$: Modular Arithmetic

This chapter is a diversion from our journey from the natural numbers to the real numbers. This side trip offers another way to use an equivalence relation and its associated partition into equivalence classes to yield a new family of numbers.

We present modular arithmetic using this motif. Informally, modular arithmetic takes a finite set of integers of the form $\mathbb{Z}_m = \{\mathbf{0}, \mathbf{1}, \mathbf{2}, \ldots, \mathbf{m-1}\}$ and operations that "wrap" around m. That is, upon reaching m, we reset to zero.

By "wrap" we mean we take the result, divide by m, and return the remainder. Thus, working with $m = 5$, we calculate the product $\mathbf{3} \cdot \mathbf{3}$ by first evaluating $3 \cdot 3$ to give 9, then divide by $m = 5$ to give a quotient of 1 (which we ignore) and a remainder of 4 yielding our answer: $\mathbf{3} \cdot \mathbf{3} = \mathbf{4}$.

For example, with $m = 5$ we have the following operation tables for addition and multiplication modulo 5:

+	0	1	2	3	4
0	0	1	2	3	4
1	1	2	3	4	0
2	2	3	4	0	1
3	3	4	0	1	2
4	4	0	1	2	3

·	0	1	2	3	4
0	0	0	0	0	0
1	0	1	2	3	4
2	0	2	4	1	3
3	0	3	1	4	2
4	0	4	3	2	1

4.1 A Relation Based on Divisibility

Let m be an integer with $m \geq 2$. We say that two integers a and b are *congruent modulo m* provided they yield the same remainder when divided by m. For example, with $m = 5$, let's see that 13 and 38 are congruent modulo 5. First we divide 13 by 5 and find the quotient (which we ignore) and remainder:

$$13 = 2 \cdot 5 + \underline{3}.$$

Likewise, we divide 38 by 5:

$$38 = 7 \cdot 5 + \underline{3}.$$

Since the remainders are both 3, the integers 13 and 38 are congruent modulo 5. Using the ideas from Proposition 2.8 this means that we have

$$a = q_1 m + r \quad \text{and} \quad b = q_2 m + r \quad \text{with } 0 \le r < m.$$

If we subtract these two equations we get $a - b = (q_1 - q_2)m$ which implies that $a - b$ is a multiple of m. This motivates us to give the following cleaner definition.

Definition 4.1 (Modular congruence)**.** Let m, a, and b be integers with $m \ge 2$. We say that *a is congruent to b modulo m* provided $m | (a - b)$.

The notation for this is $a \equiv b \pmod{m}$. Often, the number m (called the *modulus*) is clear from the context, in which case the trailing (mod m) may be omitted.

In Exercise 4.3 you are asked to show that congruence modulo m is a reflexive, symmetric, and transitive relation and thus we have the following:

Proposition 4.2. *Let m be an integer with m > 1. Congruence modulo m is an equivalence relation.*

Equivalence Classes for Congruence mod m

As we have seen, every equivalence relation has an associated partition into equivalence classes. Let's look at the case $m = 5$ starting with $[\![0]\!]$. This is the set of all integers a such that $a \equiv 0 \pmod{5}$. Those are exactly the integers divisible by 5:

$$[\![0]\!] = \{a \in \mathbb{Z} : 5 | a\} = \{\ldots, -15, -10, -5, 0, 5, 10, 15, \ldots\}.$$

Since the integer 1 is not in $[\![0]\!]$, it belongs to a different equivalence class. Note that $a \equiv 1 \pmod{5}$ means $5 | (a - 1)$, which translates to $a = 5t + 1$ for some integer t. This gives

$$[\![1]\!] = \{\ldots, -14, -9, -4, 1, 6, 11, 16, \ldots\}.$$

Likewise, we have these:

$$[\![2]\!] = \{\ldots, -13, -8, -3, 2, 7, 12, \ldots\},$$
$$[\![3]\!] = \{\ldots, -12, -7, -2, 3, 8, 13, \ldots\},$$
$$[\![4]\!] = \{\ldots, -11, -6, -1, 4, 9, 14, \ldots\}.$$

What about $[\![5]\!]$? Since $0 \equiv 5$, we see that $[\![5]\!] = [\![0]\!]$. Indeed, if we look carefully at the five classes $[\![0]\!], [\![1]\!], [\![2]\!], [\![3]\!]$, and $[\![4]\!]$, we find the every integer is in exactly one of these equivalence classes. In other words $\{[\![0]\!], [\![1]\!], [\![2]\!], [\![3]\!], [\![4]\!]\}$ is a partition of $\mathbb{Z}$.

The mod 5 case is not special in this regard. For any $m > 1$ the integers are partitioned into m equivalence classes:

$$[\![0]\!] = \{\ldots, -2m, -m, 0, m, 2m, \ldots\},$$
$$[\![1]\!] = \{\ldots, -2m+1, -m+1, 1, m+1, 2m+1, \ldots\},$$
$$[\![2]\!] = \{\ldots, -2m+2, -m+2, 2, m+2, 2m+2, \ldots\},$$
$$\vdots$$
$$[\![m-1]\!] = \{\ldots, -m-1, -1, m-1, 2m-1, 3m-1, \ldots\}.$$

4.2 Modular Arithmetic

Given an integer $m > 1$, the equivalence relation $\equiv$, congruence modulo m, partitions the integers into m equivalence classes, $[\![0]\!], [\![1]\!], \ldots, [\![m-1]\!]$. These form a set of new numbers we denote $\mathbb{Z}_m$. We write **0** for $[\![0]\!]$, **1** for $[\![1]\!]$, and so forth to give this:

$$\mathbb{Z}_m = \{\mathbf{0}, \mathbf{1}, \ldots, \mathbf{m-1}\}.$$

We imbue this set with the operations $+$ and $\cdot$ as follows.

Definition 4.3 (Arithmetic modulo m). Let $m \in \mathbb{Z}$ with $m > 1$. Let **x** and **y** be elements of $\mathbb{Z}_m$. Let x be a representative of **x** and y be a representative of **y**. Addition and multiplication in $\mathbb{Z}_m$ are defined as follows.

- $\mathbf{x} + \mathbf{y} = [\![x + y]\!]$.
- $\mathbf{x} \cdot \mathbf{y} = [\![x \cdot y]\!]$.

For example, let's take $m = 10$, $\mathbf{x} = \mathbf{6}$, and $\mathbf{y} = \mathbf{8}$. To calculate $\mathbf{6} + \mathbf{8}$ we take any representative we like from these classes. The simplest are, of course, 6 and 8. Then we have

$$\mathbf{6} + \mathbf{8} = [\![6+8]\!] = [\![14]\!] = \mathbf{4},$$
$$\mathbf{6} \cdot \mathbf{8} = [\![6 \cdot 8]\!] = [\![48]\!] = \mathbf{8}.$$

For this definition to be legitimate, the result should not depend on the choice of representatives. For this example, we could have chosen different representatives, say $-4 \in \mathbf{6}$ and $28 \in \mathbf{8}$:

$$\mathbf{6} + \mathbf{8} = [\![-4 + 28]\!] = [\![24]\!] = \mathbf{4},$$
$$\mathbf{6} \cdot \mathbf{8} = [\![-4 \cdot 28]\!] = [\![-112]\!] = \mathbf{8}.$$

Exercise 4.5 asks you to verify the legitimacy of Definition 4.3 by showing that the results of addition and multiplication do not depend on the choice of representatives.

$\mathbb{Z}_m$ is a Commutative Ring

Proposition 3.10 asserts a suite of properties that hold for addition and multiplication of integers. Specifically, both addition and multiplication are commutative, associative, and have identity elements (0 for addition and 1 for multiplication). In addition, every integer has an additive inverse. Finally, we know that multiplication distributes over addition. In other words, as we discussed earlier in connection with Proposition 3.10, $(\mathbb{Z}, +, \cdot)$ is a commutative ring.

The integers modulo m also satisfy these properties.

> **Proposition 4.4.** *Let $m > 1$ be an integer. Then $(\mathbb{Z}_m, +, \cdot)$ is a commutative ring.*

The properties are not difficult to verify. Let **x**, **y**, and **z** be elements of $\mathbb{Z}_m$ and let x, y, and z be representatives for each. Here are the calculations that verify Proposition 4.4. In each case, we rely heavily on the arithmetic properties of ordinary integers.

- Commutative:
$$\mathbf{x} + \mathbf{y} = [\![x + y]\!] = [\![y + x]\!] = \mathbf{y} + \mathbf{x},$$
$$\mathbf{x} \cdot \mathbf{y} = [\![x \cdot y]\!] = [\![y \cdot x]\!] = \mathbf{y} \cdot \mathbf{x}.$$

- Associative:
$$(\mathbf{x} + \mathbf{y}) + \mathbf{z} = [\![x + y]\!] + [\![z]\!] = [\![(x + y) + z]\!]$$
$$= [\![x + (y + z)]\!] = [\![x]\!] + [\![y + z]\!]$$
$$= \mathbf{x} + (\mathbf{y} + \mathbf{z}).$$

The calculations for multiplication are analogous.

- Identity elements:
$$\mathbf{x} + \mathbf{0} = [\![x + 0]\!] = [\![x]\!] = \mathbf{x},$$
$$\mathbf{x} \cdot \mathbf{1} = [\![x \cdot 1]\!] = [\![x]\!] = \mathbf{x}.$$

Furthermore $1 \notin \mathbf{0}$ because 1 is not divisible by m (because $m > 1$). Therefore $\mathbf{0} \neq \mathbf{1}$.

- Additive inverses:
$$\mathbf{x} + [\![-x]\!] = [\![x + (-x)]\!] = [\![0]\!] = \mathbf{0}.$$

- Distributive:

$$\begin{aligned}\mathbf{x} \cdot (\mathbf{y} + \mathbf{z}) &= [\![x]\!] \cdot [\![y+z]\!] \\ &= [\![x \cdot (y+z)]\!] = [\![x \cdot y + x \cdot z]\!] \\ &= [\![x \cdot y]\!] + [\![x \cdot z]\!] = \mathbf{x} \cdot \mathbf{y} + \mathbf{x} \cdot \mathbf{z}.\end{aligned}$$

The integers are ordered by the $\leq$ relation. Is there an order relation for $\mathbb{Z}_m$? One might be tempted to suggest this order:

$$0 \leq 1 \leq 2 \leq \cdots \leq m - 1. \qquad (*)$$

The issue here is that this order does not "play nicely" with addition and multiplication. For example, for integers $x \leq y$ implies that $x + z \leq y + z$. Using the proposed order with $m = 10$, we would have $\mathbf{5} \leq \mathbf{8}$. However, adding $\mathbf{4}$ to both sides would suggest that $\mathbf{5} + \mathbf{4} \leq \mathbf{8} + \mathbf{4}$ yielding $\mathbf{9} \leq \mathbf{2}$, which is at odds with $(*)$.

There simply isn't a way to give the elements of $\mathbb{Z}_m$ an ordering that yields a result akin to Proposition 3.12.

Recap

For each integer $m > 1$ we defined the relation *congruence modulo m* on the integers. The equivalence classes of this relation gives rise to a new set of numbers denoted $\mathbb{Z}_m$. Further, we defined addition and multiplication for $\mathbb{Z}_m$ yielding a commutative ring. However, no reasonable ordering relation is associated with these types of numbers.

Exercises

4.1 Find the additive and multiplicative inverses of $\mathbf{2}$ and of $\mathbf{6}$ in $\mathbb{Z}_9$.

4.2 Give an example of an integer $m \geq 2$ and nonzero elements $\mathbf{a}, \mathbf{b} \in \mathbb{Z}_m$ with $\mathbf{a} \cdot \mathbf{b} = \mathbf{0}$. We call such numbers *zero divisors*.

4.3 Show that $\equiv$, congruence modulo m, is reflexive, symmetric, and transitive, and hence is an equivalence relation.

4.4 The equivalence classes modulo 2 have familiar names. What are they?

4.5 Definition 4.3 tells us how to add and multiply in $\mathbb{Z}_m$. Show that the definition does not depend on the choice of representatives.

4.6 In this chapter we required $m > 1$. What's the problem with the modulus $m = 1$?

4.7 In this exercise we explore the question: For which prime numbers p (with $p > 2$) does $\mathbb{Z}_p$ have a square root of -1?

More specifically, let p be a prime. In some cases $\mathbb{Z}_p$ contains an element **a** with $\mathbf{a}^2 = \mathbf{p} - \mathbf{1}$, and for some primes p there is no such value. Stated differently, we want to know when the equation $a^2 \equiv -1 \pmod{p}$ has a solution.

For example, in $\mathbb{Z}_{29}$ we have that $\mathbf{12}^2 = \mathbf{28}$ or, equivalently, $12^2 \equiv -1 \pmod{29}$. However, in $\mathbb{Z}_7$ there is no value **a** such that $\mathbf{a}^2 = \mathbf{6}$. We can verify this by squaring all the elements of $\mathbb{Z}_7$:

a	0	1	2	3	4	5	6
a^2	0	1	4	2	2	4	1

Notice that the value of $\mathbf{a}^2$ is never **6**.

Experiment with various primes p to develop a conjecture about when $\mathbb{Z}_p$ has a square root of -1 and when it does not.

(There is a fancy term for elements of $\mathbb{Z}_p$ that have square roots; they are called *quadratic residues*. This problem asks you to conjecture, for which primes $p > 2$ is -1 a quadratic residue in $\mathbb{Z}_p$?)

Chapter 5

$\mathbb{Q}$: Rational Numbers

We began with natural numbers. They're ideal for counting problems. They can be added and multiplied, but they can't always be subtracted. Stated differently, if a and b are natural numbers, we can't always solve the equation $a + x = b$ for a natural number x.

We solved this problem by extending natural numbers to include negative values, arriving at the integers. Now we can solve equations of the form $a+x = b$, but we can't always solve[1] equations of the form $ax = b$; in other words, division is an issue.

Our next step is to build the *rational* numbers using the integers as raw materials. You may have previously seen rational numbers introduced as fractions a/b where a and b are integers and $b \neq 0$. That doesn't exactly say what a fraction is. Let's take care of the problem using our usual tricks.

5.1 An Equivalence Relation for $\mathbb{Z} \times \mathbb{Z}^*$

The set $\mathbb{Z}$ is the set of all integers: $\mathbb{Z} = \{\ldots, -3, -2, -1, 0, 1, 2, 3, \ldots\}$. Since zero denominators are forbidden in fractions it is useful to have a notation for the nonzero integers; we use $\mathbb{Z}^*$ to be the set of nonzero integers: $\{a \in \mathbb{Z}: a \neq 0\}$.

We construct integers from natural numbers by defining an equivalence relation on pairs of natural numbers, yielding equivalence classes that are the integers. We follow the same paradigm here. Rather than beginning with the notation a/b for rational numbers, we choose instead to define an equivalence relation on $\mathbb{Z} \times \mathbb{Z}^*$; this is the set of all ordered pairs (a, b) where a and b are integers with $b \neq 0$. These ordered pairs are designed to behave exactly like fractions. To make this precise, we need to say when two such pairs are

---

[1] If our ambition is to solve *all* equations of the form $ax = b$, our hopes are doomed. When $a = 0$, the product ax is also zero no matter what value we have for x. An equation of the form $0x = 3$ is never going to have a solution. This is why division by zero is undefined.

equivalent. As you know from basic arithmetic, $\frac{a}{b} = \frac{c}{d}$ exactly when $ad = bc$. We use that as the basis for the following definition.

Definition 5.1. Let (a, b) and (c, d) be elements of $\mathbb{Z} \times \mathbb{Z}^*$. Let $\equiv$ be the relation defined by

$$(a, b) \equiv (c, d) \quad \text{exactly when} \quad ad = bc.$$

For example, we have $(2, 3) \equiv (-4, -6) \equiv (10, 15)$.

Let's check that $\equiv$ is reflexive, symmetric, and transitive:

- *Reflexive*: $(a, b) \equiv (a, b)$ because $ab = ba$.

- *Symmetric*: Suppose $(a, b) \equiv (c, d)$. Then $ad = bc$ which we can rearrange as $cb = da$ giving $(c, d) \equiv (a, b)$.

- *Transitive*: Suppose $(a, b) \equiv (c, d)$ and $(c, d) \equiv (e, f)$. We want to show that $(a, b) \equiv (e, f)$.

  From what we are given we have $ad = bc$ and $cf = de$. Multiply $ad = bc$ by f to get $adf = bcf$. Multiply $cf = de$ by b to get $bcf = bde$. Since adf and bde both equal bcf, they are equal: $adf = bde$. Since $d \ne 0$ we can cancel it from both sides to give $af = be$. (See Exercise 3.13 to justify the cancellation of d from both sides.) Therefore, $(a, b) \equiv (e, f)$.

Therefore, the relation $\equiv$ from Definition 5.1 is an equivalence relation.

Definition 5.2. The *rational numbers* are the equivalence classes of the relation $\equiv$ (from Definition 5.1) on the set $\mathbb{Z} \times \mathbb{Z}^*$.

The set of all rational numbers is denoted $\mathbb{Q}$.

As usual, we use double brackets to stand for equivalence classes. Thus $[\![(2, 3)]\!]$ is a rational number that might be more comfortably written as $\frac{2}{3}$. That $(2, 3) \equiv (-4, -6) \equiv (10, 15)$ implies that $[\![(2, 3)]\!] = [\![(-4, -6)]\!] = [\![(10, 15)]\!]$, which is an elaborate way to write

$$\frac{2}{3} = \frac{-4}{-6} = \frac{10}{15}.$$

Nevertheless, we adhere to the $[\![(a, b)]\!]$ notation for the time being.

5.2 Rational Arithmetic

Having defined rational numbers, we now present definitions for addition and multiplication.

Definition 5.3. Let **x** and **y** be rational numbers and let (a, b) be a representative of **x** and (c, d) be a representative of **y**. Then

$$\mathbf{x} + \mathbf{y} = [\![(ad + bc, bd)]\!] \quad \text{and} \quad \mathbf{x} \cdot \mathbf{y} = [\![(ac, bd)]\!].$$

Note: This definition is equivalent to the usual rules for adding and multiplying fractions:

$$\frac{a}{b} + \frac{c}{d} = \frac{ad + bc}{bd} \quad \text{and} \quad \frac{a}{b} \cdot \frac{c}{d} = \frac{ac}{bd}.$$

As we have done before, we must check that the result of adding and multiplying rational numbers does not depend on the choice of representatives. See Exercise 5.4.

The addition and multiplication operations for rational numbers exhibit all of the same properties as the same-named operations for integers. Specifically $(\mathbb{Q}, +, \cdot)$ forms a commutative ring.

Let's review what this means:

- Addition and multiplication are commutative:

$$[\![(a, b)]\!] + [\![(c, d)]\!] = [\![(ad + bc, bd)]\!] = [\![(cb + da, db)]\!]$$
$$= [\![(c, d)]\!] + [\![(a, b)]\!],$$

$$[\![(a, b)]\!] \cdot [\![(c, d)]\!] = [\![(ac, bd)]\!] = [\![(ca, db)]\!] = [\![(c, d)]\!] \cdot [\![(a, b)]\!].$$

- Addition and multiplication are associative: See Exercise 5.5.

- Identity elements: Let $\mathbf{0} = [\![(0, 1)]\!]$ and $\mathbf{1} = [\![(1, 1)]\!]$. For $\mathbf{x} = [\![(a, b)]\!]$ we have:

$$\mathbf{x} + \mathbf{0} = [\![(a, b)]\!] + [\![(0, 1)]\!] = [\![(a \cdot 1 + b \cdot 0, b \cdot 1)]\!] = [\![(a, b)]\!] = \mathbf{x}$$
$$\text{and} \quad \mathbf{x} \cdot \mathbf{1} = [\![(a, b)]\!] \cdot [\![(1, 1)]\!] = [\![(a \cdot 1, b \cdot 1)]\!] = [\![(a, b)]\!] = \mathbf{x}.$$

Also note that $\mathbf{0} \neq \mathbf{1}$ because $0 \cdot 1 \neq 1 \cdot 1$.

- Additive inverses: Let $\mathbf{x} = [\![(a, b)]\!]$ and define $-\mathbf{x}$ to be $[\![(-a, b)]\!]$. Then

$$\mathbf{x} + -\mathbf{x} = [\![(a, b)]\!] + [\![(-a, b)]\!] = [\![(ab + b(-a), b^2)]\!] = [\![(0, b^2)]\!] = \mathbf{0}.$$

From this we define subtraction: $\mathbf{x} - \mathbf{y}$ is simply $\mathbf{x} + (-\mathbf{y})$.

- Distributive: See Exercise 5.6.

The rational numbers have one additional, important property. Not only do rational numbers have additive inverses, but they also (mostly) have multiplicative inverses. That is, if **x** is a nonzero rational number, there is a rational number **y** with $\mathbf{x} \cdot \mathbf{y} = \mathbf{1}$. This is easy to check. If **x** is represented by (a, b) then not only is $b \neq 0$, but also $a \neq 0$ because $\mathbf{x} \neq \mathbf{0}$. Let $\mathbf{y} = [\![(b, a)]\!]$ and we calculate:

$$\mathbf{x} \cdot \mathbf{y} = [\![(a,b)]\!] \cdot [\![(b,a)]\!] = [\![(ab, ab)]\!] = [\![(1,1)]\!] = \mathbf{1}.$$

The standard notation for the multiplicative inverse of a rational number **x** is $\mathbf{x}^{-1}$.

Because nonzero elements of $\mathbb{Q}$ have multiplicative inverses, we can solve equations for the form $\mathbf{a} \cdot \mathbf{x} = \mathbf{b}$ for **x** provided $\mathbf{a} \neq \mathbf{0}$. The solution, of course, is $\mathbf{x} = \mathbf{a}^{-1} \cdot \mathbf{b}$. This may also be expressed as $\mathbf{b} \div \mathbf{a}$ or as a fraction $\mathbf{b}/\mathbf{a}$.

A commutative ring that also has multiplicative inverses for its nonzero elements is called a *field*. Here is the formal definition:

Definition 5.4. A *field* is a triple $(\mathbb{F}, +, \cdot)$ where $\mathbb{F}$ is a set and $+$ and $\cdot$ are operations defined on that set for which the following hold:

- The operations are commutative: For all $a, b \in \mathbb{F}$ we have $a+b = b+a$ and $a \cdot b = b \cdot a$.

- The operations are associative: For all $a, b, c \in \mathbb{F}$ we have $a+(b+c) = (a+b)+c$ and $a \cdot (b \cdot c) = (a \cdot b) \cdot c$.

- The operations have identity elements, usually denoted 0 and 1 (with $0 \neq 1$). That is, for every element $a \in \mathbb{F}$ we have $0 + a = a$ and $1 \cdot a = 1$.

- Every element has a + inverse and every nonzero element has a · inverse. That is, for every $a \in \mathbb{F}$ there is a $b \in \mathbb{F}$ such that $a + b = 0$. And for every $a \in \mathbb{F}$ with $a \neq 0$, there is a $b \in \mathbb{F}$ such that $a \cdot b = 1$.

- The distributive property holds: For all $a, b, c \in \mathbb{F}$ we have $a \cdot (b+c) = a \cdot b + a \cdot c$.

We can therefore assert this:

Proposition 5.5. *The rational numbers* $(\mathbb{Q}, +, \cdot)$ *are a field.*

5.3 Order Properties of the Rational Numbers

We have defined addition and multiplication for $\mathbb{Q}$; now we want to define the $\leq$ relation. This is somewhat tricky. The following is *incorrect* (see Exercise 5.9):

> **Incorrect definition of $\leq$ for** $\mathbb{Q}$: Let $\mathbf{x} = [\![(a, b)]\!]$ and $\mathbf{y} = [\![(c, d)]\!]$ be rational numbers. Then $\mathbf{x} \leq \mathbf{y}$ provided $ad \leq bc$.

Instead, let's begin by defining what it means for a rational number to be positive or negative:

> **Definition 5.6.** Let $\mathbf{x}$ be a nonzero rational number and let (a, b) be one of its representatives. We say $\mathbf{x}$ is *positive* if a and b are either both positive or both negative integers.
> Otherwise – when a and b have opposite signs – we say that $\mathbf{x}$ is *negative*.

Here's why this definition is legitimate: Let's say $(a, b) \equiv (c, d)$ are representatives of a nonzero rational number $\mathbf{x}$. Is it possible that a and b have the same sign, but c and d have different signs? Since $(a, b) \equiv (c, d)$ we know that $ad = bc$. If a and b agree in sign but c and d disagree, then one of ad or bc would be a positive integer and the other would be a negative integer, and that's impossible.

Having defined *positive* rational numbers, we define $\leq$.

> **Definition 5.7.** Let $\mathbf{x}$ and $\mathbf{y}$ be rational numbers. We define the less-than-or-equal relation by declaring $\mathbf{x} \leq \mathbf{y}$ to be true provided $\mathbf{y} - \mathbf{x}$ is either zero or positive.

Note: This definition depends on the subtraction operation; see Exercise 5.3.

As usual, we define $<$, $\geq$, and $>$ based on $\leq$; see the boxed comments on page 38. When we say that $\mathbf{x}$ is *positive*, that is equivalent to $\mathbf{x} > \mathbf{0}$.

We have the usual properties:

> **Proposition 5.8.** *Let* $\mathbf{x}$*,* $\mathbf{y}$*, and* $\mathbf{z}$ *be rational numbers. The following properties hold.*
>
> - *Reflexive:* $\mathbf{x} \leq \mathbf{x}$.

> - *Antisymmetric:* If $\mathbf{x} \le \mathbf{y}$ and $\mathbf{y} \le \mathbf{x}$, then $\mathbf{x} = \mathbf{y}$.
> - *Transitive:* If $\mathbf{x} \le \mathbf{y}$ and $\mathbf{y} \le \mathbf{z}$, then $\mathbf{x} \le \mathbf{z}$.
> - *Trichotomy: Exactly one of the following is true:* $\mathbf{x} < \mathbf{y}$, $\mathbf{x} = \mathbf{y}$, or $\mathbf{x} > \mathbf{y}$.

See Exercise 5.10.

> **Proposition 5.9.** *Let* $\mathbf{x}$ *and* $\mathbf{y}$ *be rational numbers. If* $\mathbf{x} > \mathbf{0}$ *and* $\mathbf{y} > \mathbf{0}$ *then* $\mathbf{x} + \mathbf{y} > \mathbf{0}$ *and* $\mathbf{x} \cdot \mathbf{y} > \mathbf{0}$.

See Exercise 5.11.

The rational numbers form a field that has an ordering relation $\le$ satisfying Propositions 5.8 and 5.9; as a result we call the system $(\mathbb{Q}, +, \cdot, \le)$ an *ordered field*.

We have reached a good place. Given any rational numbers $\mathbf{a}$, $\mathbf{b}$, and $\mathbf{c}$ we can solve the equation $\mathbf{a} \cdot \mathbf{x} + \mathbf{b} = \mathbf{c}$ so long as $\mathbf{a} \ne \mathbf{0}$; the solution, of course, is $\mathbf{x} = \mathbf{a}^{-1} \cdot (\mathbf{c} - \mathbf{b})$.

5.4 Assimilation

The rational numbers of the form $[\![(a, 1)]\!]$ are an exact replica of the integers. To be more specific, let $\mathbf{a} = [\![(a, 1)]\!]$, $\mathbf{b} = [\![(b, 1)]\!]$, and $\mathbf{c} = [\![(c, 1)]\!]$.

$$\begin{array}{ccc} \mathbf{a} + \mathbf{b} = \mathbf{c} & \leftrightarrow & a + b = c, \\ \mathbf{a} \cdot \mathbf{b} = \mathbf{c} & \leftrightarrow & a \cdot b = c, \\ \mathbf{a} \le \mathbf{b} & \leftrightarrow & a \le b. \end{array}$$

In this way, the rational numbers, $\mathbb{Q}$, contain an exact replica of the integers, $\mathbb{Z}$. Thus, just as in Section 3.5 we considered $\mathbb{N}$ to be a subset of $\mathbb{Z}$, going forward we find the integers as simply a subset of $\mathbb{Q}$. We now have $\mathbb{N} \subseteq \mathbb{Z} \subseteq \mathbb{Q}$. Next stop: the real numbers!

Recap

We defined an equivalence relation on $\mathbb{Z} \times \mathbb{Z}^*$ so that ordered pairs (a, b) behave like fractions a/b. The equivalence classes of this relation are the rational numbers $\mathbb{Q}$. We defined addition and multiplication of rational numbers, and then noted that $(\mathbb{Q}, +, \cdot, \le)$ form an ordered field.

Exercises

5.1 Let a and b be integers with $b \neq 0$. Which of the following must be true:

   a) $(a, b) \equiv (-a, -b)$.
   b) $(a, -b) \equiv (-a, b)$.
   c) $(a, b) \equiv (b, a)$.

5.2 Let a and b be integers with $b \neq 0$. Evaluate $[\![(a, b)]\!] + [\![(a, b)]\!]$ and $[\![(a, b)]\!] \cdot [\![(a, b)]\!]$.

5.3 Let **x** and **y** be integers. Use representatives of **x** and **y** to define **x** − **y** and, assuming **y** ≠ **0**, to define **x** ÷ **y**.

5.4 Show that the result of adding and multiplying rational numbers does not depend on the choice of representatives in Definition 5.3.

   Specifically, show that if $(a, b) \equiv (a', b')$ and $(c, d) \equiv (c', d')$ then

   $$(ad + bc, bd) \equiv (a'd' + b'c', b'd') \quad \text{and} \quad (ac, bd) \equiv (a'c', b'd').$$

5.5 Show that for rational numbers the operations addition and multiplication are associative.

5.6 Show that for rational numbers **x**, **y**, and **z** we have $\mathbf{x} \cdot (\mathbf{y} + \mathbf{z}) = \mathbf{x} \cdot \mathbf{y} + \mathbf{x} \cdot \mathbf{z}$.

5.7 In Chapter 4 we showed that $(\mathbb{Z}_m, +, \cdot)$ is a commutative ring. For some values of m it is also a field, and for some it is not. For example, when $m = 5$ (see the tables on page 67) we see that every nonzero element has a multiplicative inverse: $1^{-1} = 1$, $2^{-1} = 3$, $3^{-1} = 2$, and $4^{-1} = 4$.

   Show that $(\mathbb{Z}_6, +, \cdot)$ is not a field, but $(\mathbb{Z}_7, +, \cdot)$ is. Make a conjecture for when $(\mathbb{Z}_m, +, \cdot)$ is a field.

5.8 In the previous problem we noted that $(\mathbb{Z}_5, +, \cdot)$ is a field. Suppose we define $\leq$ for $\mathbb{Z}_5$ so that we have $0 < 1 < 2 < 3 < 4$. Is $(\mathbb{Z}_5, +, \cdot, \leq)$ an ordered field?

5.9 We said that the attempted definition for $\leq$ on page 77 is incorrect. Why?

5.10 Verify Proposition 5.8.

5.11 Verify Proposition 5.9.

5.12 The rational numbers form a *dense* set. This means that, given two different rational numbers, there is another rational number strictly between them.

   If **x** < **y** are rational numbers, find a rational number **z** with **x** < **z** < **y**.

5.13 In Exercise 3.18 we asked if the set of all polynomials with integer coefficients, $\mathbb{Z}[x]$, forms a commutative ring. Just as rational numbers are fractions with integer numerators and denominators, a *rational function* is a fraction whose numerator and denominator are polynomials.

Let $\mathbb{Z}(x)$ stand for the set of all such functions with integer coefficients. That is, $\mathbb{Z}(x)$ consists of all functions of the form $p(x)/q(x)$ were $p, q \in \mathbb{Z}[x]$ and q is not the zero polynomial.

Is $\mathbb{Z}(x)$ a field?

Chapter 6

$\mathbb{R}$: Real Numbers I, Dedekind Cuts

In this chapter we present the first of two ways to create real numbers by building on the rational numbers, $\mathbb{Q}$. In Chapter 7 we present a different method. Each approach has its pros and cons.

The method we present in this chapter has the advantage that the definition of a real number is not too complicated, but the definitions of addition and multiplication are challenging.

In Chapter 7 the definition of a real number is more complicated, but addition and multiplication are simpler.

Wait. There are two different definitions of real numbers?

That's right. We are giving two different approaches, and neither definition involves decimals. This tension is resolved in Chapter 8 in which we explain that, while the definitions are superficially different, they give rise to the same essential structure that we call *the* real number system.

6.1 The Case for Real Numbers

Before we start, we ask: Why do we need real numbers anyhow? From the perspective of applications of mathematics, it would be exceedingly rare to need numbers with, say, more than 20 digits of accuracy. Indeed, that level of precision exceeds what is typically found in most calculators and computer programming languages. Suppose we assume

$$\pi = \frac{31415926535897932384626433832795028841971}{10000000000000000000000000000000000000000} \in \mathbb{Q}.$$

Is this going to lead to an error in any applied setting? Almost certainly not. And, if in some strange case we need something more accurate, we could just grab more digits.

However, from the perspective of mathematics, the rational numbers are insufficient, and we illustrate this with a classical example. Suppose we have a 1×1 square as illustrated in Figure 6.1. What is the length of its diagonal?

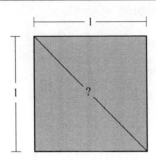

Figure 6.1: What *exactly* is the length of the diagonal of a one-by-one square?

We know how to find the answer. Let x be the length of the diagonal. By the Pythagorean Theorem we have

$$1^2 + 1^2 = x^2,$$

from which we derive $x = \sqrt{2}$.

What is the square root of 2? One answer: It's a number that when multiplied by itself gives 2. This is an extremely unhelpful answer.

Let's do just a bit of calculating: Notice that $1.4^2 = 1.96$ and $1.5^2 = 2.25$. We infer that $1.4 < \sqrt{2} < 1.5$, but that doesn't answer the question. A single digit to the right of the decimal point doesn't give us much, so consider this:

$$1.414213^2 \approx 1.99999841,$$
$$1.414214^2 \approx 2.00000124.$$

and therefore $1.414213 < \sqrt{2} < 1.414214$. Alas, this still doesn't tell us exactly what $\sqrt{2}$ is.

If there were a rational number q such that $q^2 = 2$, that would be a satisfying, exact answer to the question. However, as you likely know, there is no such number.

Proposition 6.1. *Let $x \in \mathbb{Q}$. Then $x^2 \neq 2$.*

Proof. For sake of contradiction, let $x \in \mathbb{Q}$ with $x^2 = 2$. Since x is a rational number it equals b/a where a and b are positive integers. Therefore $(b/a)^2 = 2$.

We can rewrite this as
$$b^2 = 2a^2. \qquad (*)$$
The left and right sides of $(*)$ represent the same integer, N. By the Fundamental Theorem of Arithmetic (Theorem 2.10) N has a unique factorization into primes.

How many times does the prime 2 appear in the factorization of N?

Since $N = b^2$, the number of times that 2 is a factor of N is exactly double the number of times 2 is a factor of b. Thus, in N's factorization the prime number 2 appears an *even* number of times.

Similarly, $N = 2a^2$. In this case, 2 appears a certain number of times in the factorization of a. The number of times 2 appears in the factorization of a^2 is exactly double that amount. Finally, there's one additional factor of 2 comprising N (because $N = 2 \cdot a \cdot a$). Therefore 2 appears an *odd* number of times in the factorization of N.

This is a contradiction and therefore there is no rational number x such that $x^2 = 2$. □

There is no solution to the equation $x^2 = 2$ in the rational numbers. If there is a $\sqrt{2}$, it's difficult to say *exactly* what it is.

6.2 Left Rays and Real Numbers

The definition of a real number we present in this chapter is essentially the one given by the nineteen-century German mathematician Richard Dedekind. Just as we use natural numbers as the raw ingredients to create integers, and integers to create rational numbers, we use rational numbers to build real numbers.

Dedekind imagined the rational numbers as a number line, and defined real numbers as "cuts" that partition $\mathbb{Q}$ into a left half and right half. We use that core idea in our approach in this chapter.

Definition 6.2. A *left ray* is a set L of rational numbers with the properties:

- $L \neq \emptyset$ and $L \neq \mathbb{Q}$,

- if $a < b$ and $b \in L$ then $a \in L$.

In other words, a *left ray* is a nonempty, proper subset of the rationals with the property that it goes "forever to the left." However, because a left ray does not include all rationals, it does not continue "forever to the right."

Here are two examples of left rays:

$$L = \{a \in \mathbb{Q} : a < 0\} \quad \text{and} \quad L' = \{a \in \mathbb{Q} : a \leq 0\}.$$

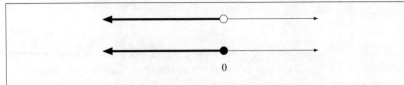

Figure 6.2: An illustration of equivalent left rays $L = \{a \in \mathbb{Q}: a < 0\}$ (top) and $L' = \{a \in \mathbb{Q}: a \leq 0\}$ (bottom).

These two left rays are illustrated in Figure 6.2. The upper drawing represents L because the open circle means that 0 is missing from the set, but all negative rational numbers are included. The lower drawing represents L' because the filled circle means that 0 is included.

When a left ray, such as L, does not contain a maximum value we call that ray *open*. Left rays that include a maximum value, such as L', are called *closed*.

These two sets are candidates for representing the real number **0**. Which shall we use? Both! To that end, we define an equivalence relation for left rays.

Definition 6.3. We say that left rays L and L' are equivalent provided L and L' are the same or differ in exactly one element.

Stated differently, $L \equiv L'$ provided that the symmetric difference $L \triangle L'$ is either empty or a singleton.

(See Exercise 1.13 for the definition of symmetric difference of sets.)

The two left rays illustrated in Figure 6.2 are equivalent because the only difference between the sets is whether or not 0 is included. In symbols:

$$L \triangle L' = \{a \in \mathbb{Q}: a < 0\} \triangle \{a \in \mathbb{Q}: a \leq 0\} = \{0\}.$$

Notice that different, but equivalent, left rays differ only in whether or not they have a right end point: L does not have a right end point but L' does. So the equivalence class of L (or of L') is just the set $[\![L]\!] = \{L, L'\}$.

Now let's think about a different left ray:

$$R = \{x \in \mathbb{Q}: x < 0\} \cup \{x \in \mathbb{Q}: x^2 < 2\}.$$

This set is defined by taking the union of two pieces: all negative rational numbers and all rational numbers whose square is less than 2. See Figure 6.3.

Let's check that R is a left ray by consulting Definition 6.2.

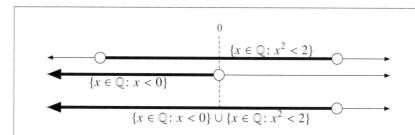

Figure 6.3: This is an illustration of the set $R = \{x \in \mathbb{Q}: x < 0\} \cup \{x \in \mathbb{Q}: x^2 < 2\}$. Notice that R is a left ray.

First we need to see that R is a nonempty proper subset of $\mathbb{Q}$. It's nonempty because $-1 \in R$. It is also not all of $\mathbb{Q}$ because 3 is not negative (so not in the first piece) nor does 3 satisfy the condition $x^2 < 2$. Therefore $3 \notin R$.

Next we need to check the "leftward" condition: if $a < b$ and $b \in R$, then we must have $a \in R$.

- If b is negative then, since $a < b$, a is also negative. Therefore $a \in R$.
- If b is nonnegative, then $b^2 < 2$. If a is negative, then we know $a \in R$, so suppose a is also nonnegative. In this case we have $0 \le a < b$ which implies $a^2 < b^2 < 2$. Therefore $a \in R$.

In either case $a < b$ and $b \in R$ imply $a \in R$. Therefore R is a left ray. (This is also easily understood by looking at the picture in Figure 6.3.)

Is there a different left ray R' that is equivalent to R? That is, can we "fill in" the empty spot at the far right? Perhaps this:

$$R' = \{x \in \mathbb{Q}: x < 0\} \cup \{x \in \mathbb{Q}: x^2 \le 2\}.$$

Notice that we replaced $x^2 < 2$ with $x^2 \le 2$. But this doesn't change anything because, as we have shown, there is no rational number x with $x^2 = 2$.

Therefore, for this left ray, we have $[\![R]\!] = \{R\}$. There are no other left rays equivalent to R.

Having defined left rays and equivalence of left rays, we are ready to give the definition of a real number.

Definition 6.4. A *real number* is an equivalence class of left rays. The set of all real numbers is denoted $\mathbb{R}$.

Recalling the two examples from this section, the equivalence class $[\![L]\!] = \{L, L'\}$, where

$$L = \{a \in \mathbb{Q} : a < 0\} \quad \text{and} \quad L' = \{a \in \mathbb{Q} : a \leq 0\},$$

is the real number **0**.

The equivalence class $[\![R]\!] = \{R\}$, where

$$R = \{x \in \mathbb{Q} : x < 0\} \cup \{x \in \mathbb{Q} : x^2 < 2\},$$

is the real number $\sqrt{2}$.

For this all to work out, we need to explain how to add, multiply, and compare real numbers.

6.3 Addition

Addition of real numbers rests on addition of rational numbers. Roughly speaking, we take all the numbers in one left ray, add them to all the numbers in another left ray, and the result is a new left ray that represents the sum. Let's make this more precise.

Suppose we have two left rays, L_1 and L_2. We can form a new left ray simply by collecting all possible sums $x + y$ where $x \in L_1$ and $y \in L_2$. That is, we form the set

$$L = \{x + y : x \in L_1 \text{ and } y \in L_2\}.$$

The result is again a left ray. See Exercise 6.2. Let's use the notation $L = L_1 + L_2$.

For example, suppose $L_1 = \{x \in \mathbb{Q} : x < 1\}$ and $L_2 = \{x \in \mathbb{Q} : x < 2\}$. Then $L_1 + L_2$ is the set formed by adding all rational numbers less than 1 to all rational numbers less than 2. The result is $L_1 + L_2 = \{x \in \mathbb{Q} : x < 3\}$.

Some left rays have a largest element. Here L_1 and L_2 are equivalent to the sets

$$L_1' = \{x \in \mathbb{Q} : x \leq 1\} \equiv L_1 \quad \text{and} \quad L_2' = \{x \in \mathbb{Q} : x \leq 2\} \equiv L_2.$$

Their sum is

$$L_1' + L_2' = \{x \in \mathbb{Q} : x \leq 3\} \equiv L_1 + L_2.$$

This inspires the following definition.

Definition 6.5. Let **a** and **b** be real numbers. Let L be a representative of **a** and R be a representative of **b**. The sum **a** + **b** is defined to be $[\![L + R]\!]$.

Left Rays and Decimals

At first glance, equivalence classes of left rays and decimals appear to be completely different, but there is a natural connection that we can illustrate with the real number **5/4**. There are two different decimal representations for this number, namely

$$1.2500000\ldots \quad \text{and} \quad 1.2499999\ldots.$$

Each of these can be converted into a left ray by taking more and more digits from the decimal expression. The $2.50000\ldots$ representation converts like this:

$$\begin{aligned}
L = &\{x \in \mathbb{Q}: x \leq 1\} \\
&\cup \{x \in \mathbb{Q}: x \leq 1.2\} \\
&\cup \{x \in \mathbb{Q}: x \leq 1.25\} \\
&\cup \{x \in \mathbb{Q}: x \leq 1.250\} \\
&\cup \{x \in \mathbb{Q}: x \leq 1.2500\} \cup \cdots \\
= &\{x \in \mathbb{Q}: x \leq 5/4\}.
\end{aligned}$$

This is a closed left ray.

The $1.249999\ldots$ expansion is more interesting; it converts like this:

$$\begin{aligned}
L' = &\{x \in \mathbb{Q}: x \leq 1\} \\
&\cup \{x \in \mathbb{Q}: x \leq 1.2\} \\
&\cup \{x \in \mathbb{Q}: x \leq 1.24\} \\
&\cup \{x \in \mathbb{Q}: x \leq 1.249\} \\
&\cup \{x \in \mathbb{Q}: x \leq 1.2499\} \cup \cdots \\
= &\{x \in \mathbb{Q}: x < 5/4\}
\end{aligned}$$

This is an open left ray, so it is not the same as L. However, they are equivalent left rays, so they are both representatives of the real number **5/4** $= \{L, L'\}$.

Exercise 6.3 illustrates that the choice of representative in Definition 6.3 does not affect the result. That is, if L and L' are representatives of **a**, and if R and R' are representatives of **b**, then all of $L + R$, $L + R'$, $L' + R$, and $L' + R'$ are equivalent and represent **a** + **b**.

The commutative and associative properties for addition of real numbers follow directly from the analogous properties for rational numbers.

For example, if L and L' are left rays, we see that

$$L + L' = \{a + b : a \in L \text{ and } a \in L'\} = \{b + a : b \in L' \text{ and } a \in L\} = L' + L.$$

Therefore, for real numbers **x** and **y** we have $\mathbf{x} + \mathbf{y} = \mathbf{y} + \mathbf{x}$.

Identity Element for Addition

The real number **0** is $[\![\{x \in \mathbb{Q} : x < 0\}]\!] = [\![\{x \in \mathbb{Q} : x \le 0\}]\!]$. What happens when we add another real number, **a** to **0**?

To understand this, let L be a left ray. We'll show that $L + \{x \in \mathbb{Q} : x \le 0\} = L$. (The argument that $L + \{x \in \mathbb{Q} : x < 0\} = L$ is a bit more complicated, but has a similar flavor.)

To show that two sets are the same, we demonstrate the every element in one set is in the other.

First, suppose that $a \in L$. Since $0 \in \{x \in \mathbb{Q} : x \le 0\}$ we know that $a + 0 \in L + \{x \in \mathbb{Q} : x \le 0\}$; that is, $a \in \{x \in \mathbb{Q} : x \le 0\}$.

Second, suppose that $a \in L + \{x \in \mathbb{Q} : x \le 0\}$. This means that $a = b + c$ where $b \in L$ and $c \in \{x \in \mathbb{Q} : x \le 0\}$. Since $c \le 0$ we know that $a = b + c \le b$, and we know that $b \in L$. Since L is a left ray, $a < b \in L$ implies that $a \in L$.

Therefore $L + \{x \in \mathbb{Q} : x \le 0\} = L$. Therefore, for any real number **x**, we have that $\mathbf{x} + \mathbf{0} = \mathbf{x}$.

Additive Inverses

Given a real number **x**, we need to define $-\mathbf{x}$. A reasonable – but incorrect! – idea is to take a left ray representing **x** and negate all of its elements. That is, if L is a left ray, let

$$-L = \{-a : a \in L\}. \quad \leftarrow \text{This doesn't work!!}$$

The problem is that this doesn't create a left ray. For example, let L be a representative of the real number **1**, say $L = \{x \in \mathbb{Q} : x \le 1\}$. Notice that all the negative rational numbers in L become positive when they are negated, and the rational numbers from 0 up to 1 become the numbers from -1 to 0.

By this definition, $-L = \{x \in \mathbb{Q} : x \ge -1\}$. What we see is that when we negate all the elements of a left ray, we get a right ray.

We know that a left ray representing $-\mathbf{1}$ should be either $\{x \in \mathbb{Q} : x < -1\}$ or $\{x \in \mathbb{Q} : x \le -1\}$. We can salvage the faulty definition for $-L$ by realizing that the left ray we want is all the rational numbers *not* in $-L$.

The negation of a real number is constructed in two steps:

- A "flip" step where the left ray is flipped around 0. The left ray L becomes a right ray R.

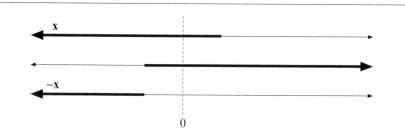

Figure 6.4: A geometric construction of the negative of a real number. The first step is to replace each element of the left ray with its negative: this is a flip around 0 creating a right ray. The second step is to take the complement of the set created in the first step.

- A "swap" step in which we form the complement of R, namely, $\{x \in \mathbb{Q} : x \notin R\}$.

This flip-and-swap operation is illustrated in Figure 6.4.

Notice that when we flip-and-swap a closed left ray, the result will be an open left ray. A flip-and-swap of an open left ray might yield a closed left ray or an open left ray. See Exercise 6.6. Of course, this doesn't matter since the presence or absence of the end point of a left ray yields an equivalent representation of the real number.

The final task we face is to show that $\mathbf{x} + (-\mathbf{x}) = \mathbf{0}$. To accomplish this algebraically is annoyingly complicated because the swap step in the construction is not easily expressed by a formula. Instead, we settle for an informal geometric explanation.

Let L be an open left ray representing a real number $\mathbf{x}$ and let L' be an open left ray representing $-\mathbf{x}$.

Notice that the (missing) right end points of the rays L and L' are equidistant from 0. Hence, if we choose $a \in L$ and $b \in L'$ we see that $a + b < 0$. This implies that $L + L'$ contains only negative rational numbers. That is, $L + L' \subseteq \{x \in \mathbb{Q} : x < 0\}$.

On the other hand, suppose t is a negative rational number. We want to show that $t \in L + L'$. To do this, choose a rational number $a \in L$ that is within distance $\frac{1}{3}|t|$ of the right end point of L and a point $b \in L'$ that is within distance $\frac{1}{3}|t|$ of the right end point of L'. In this way, we know that $a + b \in L + L'$.

Since a and b are each within distance $\frac{1}{3}|t|$ of their respective ray's right end point, and those right end points are symmetric around 0, we see that $a + b$ is within distance $\frac{2}{3}|t|$ of 0. That is, since $a + b$ and t are negative, we have

$a + b \geq \frac{2}{3}t$ and since $\frac{2}{3}t > t$ we have
$$t < a + b \in L + L',$$
which implies that $t \in L + L'$. Therefore $\{x \in \mathbb{Q} : x < 0\} \subseteq L + L'$.

Since we have shown that both $L + L' \subseteq \{x \in \mathbb{Q} : x < 0\}$ and $\{x \in \mathbb{Q} : x < 0\} \subseteq L + L'$ we know that $L + L' = \{x \in \mathbb{Q} : x < 0\}$ and therefore $\mathbf{x} + -\mathbf{x} = \mathbf{0}$.

We record here relevant results for addition of real numbers.

Proposition 6.6. *The real numbers, $\mathbb{R}$, and the operation of addition, $+$, satisfy the following properties:*

- *For $\mathbf{x}, \mathbf{y} \in \mathbb{R}$ we have $\mathbf{x} + \mathbf{y} = \mathbf{y} + \mathbf{x}$.*

- *For $\mathbf{x}, \mathbf{y}, \mathbf{z} \in \mathbb{R}$ we have $(\mathbf{x} + \mathbf{y}) + \mathbf{z} = \mathbf{x} + (\mathbf{y} + \mathbf{z})$.*

- *There is a real number $\mathbf{0}$ with the property that for any $\mathbf{x} \in \mathbb{R}$ we have $\mathbf{0} + \mathbf{x} = \mathbf{x}$.*

- *For every $\mathbf{x} \in \mathbb{R}$ there is a $-\mathbf{x} \in \mathbb{R}$ with the property that $\mathbf{x} + (-\mathbf{x}) = \mathbf{0}$.*

6.4 Less-than-or-equal

Addition of real numbers is complicated and, alas, multiplication considerably more complicated. Mercifully, the less-than-or-equal relation for $\mathbb{R}$ is simple.

Definition 6.7. Let $\mathbf{x}$ and $\mathbf{y}$ be real numbers represented by open left rays L and R. We have $\mathbf{x} \leq \mathbf{y}$ exactly when $L \subseteq R$.

From this we get the following.

Proposition 6.8. *The following properties hold for the $\leq$ relation for $\mathbb{R}$:*

- *For all $\mathbf{x} \in \mathbb{R}$, $\mathbf{x} \leq \mathbf{x}$. [Reflexive]*

- *For all $\mathbf{x}, \mathbf{y} \in \mathbb{R}$, if $\mathbf{x} \leq \mathbf{y}$ and $\mathbf{y} \leq \mathbf{x}$, then $\mathbf{x} = \mathbf{y}$. [Antisymmetric]*

- *For all $\mathbf{x}, \mathbf{y}, \mathbf{z} \in \mathbb{R}$, if $\mathbf{x} \leq \mathbf{y}$ and $\mathbf{y} \leq \mathbf{z}$ then $\mathbf{x} \leq \mathbf{z}$. [Transitive]*

- *For all $\mathbf{x}, \mathbf{y} \in \mathbb{R}$, exactly one of the following is true: $\mathbf{x} < \mathbf{y}$, $\mathbf{x} = \mathbf{y}$, or $\mathbf{x} > \mathbf{y}$. [Trichotomy]*

The positive real numbers are, of course, those **x** with **x > 0**.

Recall that **0** is represented by the open left ray $\{x \in \mathbb{Q}: x < 0\}$. For a real number **x** to be positive, its left ray must contain all the negative rational numbers and not "stop." It must also contain 0 and some positive rational numbers. Likewise, the closed left ray representing **x** must contain positive rational numbers.

This implies the following easy condition.

Proposition 6.9. *A real number* **x** *is positive exactly when its left ray(s) includes some positive rational numbers.*

The addition operation has the expected interactions with the $\leq$ relation, most notably this:

Proposition 6.10. *Let* **x** *and* **y** *be positive real numbers. Then* **x + y** *is positive.*

We leave the justification to you; see Exercise 6.13.

6.5 Multiplication

Starting with the Positive

Defining multiplication for real numbers is more difficult than defining addition. To calculate **2 + 3** we add all the rational numbers in a left ray for **2** to all the rational numbers in a left ray for **3**. Say,

$$L_2 = \{x \in \mathbb{Q}: x < 2\} \quad \text{and} \quad L_3 = \{x \in \mathbb{Q}: x < 3\}.$$

Then

$$L_2 + L_3 = \{x \in \mathbb{Q}: x < 5\},$$

which is a representative for **5**, as we expect.

What if we try the same idea but, instead of *adding* all the rationals in L_2 to all the rationals in L_3, we *multiply*? We get a mess! All of the negative numbers in L_2 are multiplied by all of the negative numbers in L_3 resulting in all sorts of positive numbers we do not want in the result. See Exercise 6.14.

Let's try something different with the foreknowledge that $2 \cdot 3 = 6$. The open left ray representing **6** is $L_6 = \{x \in \mathbb{Q}: x < 6\}$. Notice that if we multiply all the *nonnegative* elements of L_2 with all the *nonnegative* elements of L_3 we get the

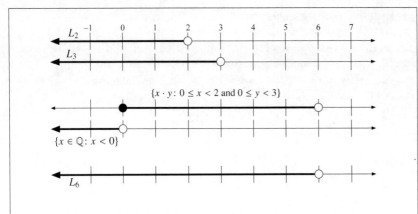

Figure 6.5: Multiplying nonnegative real numbers, **2 · 3**.

set $\{x \in \mathbb{Q} : 0 \leq x < 6\}$, which is the nonnegative part of L_6; what's missing is the long tail to the left. We can now fix this by pinning the tail on:

$$L_6 = \underbrace{\{x \cdot y : 0 \leq x < 2 \text{ and } 0 \leq y < 3\}}_{\text{nonnegative part}} \cup \underbrace{\{x \in \mathbb{Q} : x < 0\}}_{\text{tail}}.$$

This is illustrated in Figure 6.5.

This example enables us write the following definition for multiplication of nonnegative real numbers.

Definition 6.11. Let **x** and **y** be nonnegative real numbers represented by left rays L and R, respectively. Then **x · y** is the real number represented by

$$\{a \cdot b : a \in L, \ a \geq 0, \ b \in R, \text{ and } b \geq 0\} \cup \{x \in \mathbb{Q} : x < 0\}.$$

In words, to multiply nonnegative real numbers **x** and **y**, we multiply all the nonnegative elements of a representative of **x** with all the nonnegative elements of a representative of **y**, and then include all negative rationals; this gives a left ray that represents **x · y**.

Revisiting an Important Real Number

Before we extend Definition 6.11 to include negative real numbers, we recall the real number represented by

$$R = \{x \in \mathbb{Q} : x < 0\} \cup \{x \in \mathbb{Q} : x^2 < 2\}.$$

This ray is illustrated in Figure 6.3.

Let $\mathbf{a} = [\![R]\!]$ and let $\mathbf{b} = \mathbf{a}^2 = \mathbf{a} \cdot \mathbf{b}$. What is this real number $\mathbf{b}$?

To answer, we need to understand the ray L that represents $\mathbf{b}$. Using Definition 6.11 we have

$$L = \{x \in \mathbb{Q} : x < 0\} \cup \{x \cdot y : x, y \in R \text{ and } x, y \geq 0\}.$$

In other words, the "head" contains all possible products $x \cdot y$ where x and y are nonnegative elements of R.

What rational numbers are in L?

Of course all negative rationals are in L, as is 0. Now we need to determine which positive rationals are in L.

- **Claim**: *No rational number greater than or equal to 2 is in L.*

  We need to show that if $t \in L$ then $t < 2$. We know that $t = x \cdot y$ where $x, y \in R$. By the definition of R we know that $x^2 < 2$ and $y^2 < 2$. Consider t^2; we have:

  $$t^2 = (x \cdot y)^2 = x^2 \cdot y^2 < 2 \cdot 2 = 4$$

  and therefore $t < 2$.

- **Claim**: *All positive rational numbers less than 2 are in L.*

  Let t be a positive rational number with $t < 2$. We show that $t \in L$. The key insight is that there is a positive rational number s with $t < s^2 < 2$. For expediency sake, we give a "picture proof" deferring a careful, rigorous proof to the end of this chapter. (See Proposition 6.19 in which we show that strictly between any two positive rational numbers one can always find the square of a rational number.)

  For now, however, consider the graph of $y = x^2$ as illustrated in Figure 6.6. Locate t and 2 on the y-axis. There is a range of values on the x-axis whose squares lie strictly between t and 2. Let s be a rational number in that region so $t < s^2 < 2$.

  Since $s^2 < 2$, it follows that $s \in R$, and that implies that $s \cdot s = s^2 \in L$.

  Finally, since $s^2 \in L$ and $t < s^2$, it follows that $t \in L$.

We have shown that L contains exactly all rational numbers less than 2 and no others. In other words

$$L = \{x \in \mathbb{Q} : x < 2\}.$$

And since $\mathbf{2} = [\![L]\!]$ we have $\mathbf{a} \cdot \mathbf{a} = \mathbf{b} = \mathbf{2}$ or, in other words we have the following result.

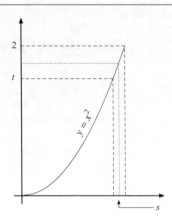

Figure 6.6: Given a rational number t with $0 < t < 2$, we find a rational number s so that $t < s^2 < 2$.

Proposition 6.12. *The real numbers contain a square root of* **2**.

Huzzah!

Multiplication for All

Definition 6.11 allows us to multiply positive real numbers and easily extends to the case where one or both factors is **0**, in which case the result is **0**. The extension to negative factors simply prescribes the properties we expect.

Definition 6.13. Let **a** and **b** be real numbers.

- If **a** and **b** are both nonnegative, use Definition 6.11 to find $\mathbf{a} \cdot \mathbf{b}$.

- If **a** is nonnegative and **b** is negative, then $\mathbf{a} \cdot \mathbf{b} = -[\mathbf{a} \cdot (-\mathbf{b})]$.

- If **a** is negative and **b** is nonnegative, then $\mathbf{a} \cdot \mathbf{b} = -[(-\mathbf{a}) \cdot \mathbf{b}]$.

- Finally, if **a** and **b** are both negative, then $\mathbf{a} \cdot \mathbf{b} = (-\mathbf{a}) \cdot (-\mathbf{b})$.

The next step is to verify the usual algebraic properties of multiplication. In each case, we begin with verifying the property for positive real numbers and then apply Definition 6.13 to check the various cases in which terms are non-positive.

For example, multiplication of real numbers is commutative. For positive reals **a** and **b** represented by left rays L and R, respectively we have

$$\mathbf{a} \cdot \mathbf{b} = [\![\{x \in \mathbb{Q}: x < 0\} \cup \{a \cdot b: a \geq 0, b \geq 0, a \in L, \text{ and } b \in R\}]\!]$$
$$= [\![\{x \in \mathbb{Q}: x < 0\} \cup \{b \cdot a: b \geq 0, a \geq 0, b \in R, \text{ and } a \in L\}]\!] = \mathbf{b} \cdot \mathbf{a}.$$

Then we just grind through the cases:

- **a** and/or **b** equal to **0**: $\mathbf{a} \cdot \mathbf{b} = \mathbf{0} = \mathbf{b} \cdot \mathbf{a}$.
- **a** positive and **b** negative: $\mathbf{a} \cdot \mathbf{b} = -[\mathbf{a} \cdot (-\mathbf{b})] = -[(-\mathbf{b}) \cdot \mathbf{a}] = \mathbf{b} \cdot \mathbf{a}$.
- **a** negative and **b** positive: $\mathbf{a} \cdot \mathbf{b} = -[(-\mathbf{a}) \cdot \mathbf{b}] = -[\mathbf{b} \cdot (-\mathbf{a})] = \mathbf{b} \cdot \mathbf{a}$.
- **a** and **b** both negative: $\mathbf{a} \cdot \mathbf{b} = (-\mathbf{a}) \cdot (-\mathbf{b}) = (-\mathbf{b}) \cdot (-\mathbf{a}) = \mathbf{b} \cdot \mathbf{a}$.

Verifying the associative property is just as uninteresting.

Multiplication of reals has an identity element

$$\mathbf{1} = [\![\{x \in \mathbb{Q}: x < 1\}]\!]$$

and it is routine to check that $\mathbf{a} \cdot \mathbf{1} = \mathbf{a}$; start by verifying this for **a** positive, and then handle $\mathbf{a} = \mathbf{0}$ and **a** negative as separate cases.

The distributive property $\mathbf{a} \cdot (\mathbf{b} + \mathbf{c}) = \mathbf{a} \cdot \mathbf{b} + \mathbf{a} \cdot \mathbf{c}$ is mildly messy in the case when all three numbers are positive. The extension to all possible real numbers is a lengthy journey through all the different positive/zero/negative possibilities.

Using Definition 6.11, if **a** and **b** are both positive, then one can check that $\mathbf{a} \cdot \mathbf{b}$ contains positive values and therefore $\mathbf{a} \cdot \mathbf{b}$ is also positive.

Multiplicative Inverses

We need to show that every nonzero real number has a multiplicative inverse.

We begin with positive numbers and use **2** as inspiration. We know that **2** is represented by the equivalent rays $\{x \in \mathbb{Q}: x < 2\}$ and $\{x \in \mathbb{Q}: x \leq 2\}$. We correctly expect its inverse, $\mathbf{2}^{-1}$, to be represented by $\{x \in \mathbb{Q}: x < \frac{1}{2}\}$ and $\{x \in \mathbb{Q}: x \leq \frac{1}{2}\}$.

Let's break this into two pieces (as we did for the multiplication in Definition 6.11). We know we need to include the "tail" consisting of all nonpositive rational numbers. Then we attach a "head" containing all the rationals between 0 and $\frac{1}{2}$:

$$\mathbf{2}^{-1} = [\![\{x \in \mathbb{Q}: x \leq 0\} \cup \{x \in \mathbb{Q}: 0 < x < \tfrac{1}{2}\}]\!].$$

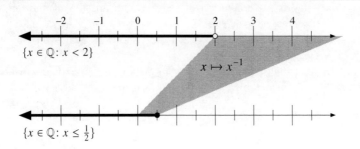

Figure 6.7: The upper portion of this figure presents the ray $L = \{x \in \mathbb{Q}: x < 2\}$ which is a representative of the real number **2**. To construct $\mathbf{2}^{-1}$ we calculate the reciprocal of every rational number not in L; that is, we include x^{-1} for all $x \notin L$, that is, for all $x \geq 2$. This is illustrated by the shaded region between the number lines. Finally, we include all $x \leq 0$ to arrive at $\{x \in \mathbb{Q}: x \leq \frac{1}{2}\}$ which is a representative of $\mathbf{2}^{-1}$.

Let's generalize this for any positive real number **a**. The "tail" part is easy: we must include $\{x \in \mathbb{Q}: x \leq 0\}$. For the "head," we start by taking the reciprocals of all positive rational numbers *not* in a representative ray L:

$$\{x^{-1}: x \in \mathbb{Q},\ x > 0,\ \text{and}\ x \notin L\}.$$

For example, with $\mathbf{a} = 2$, this set contains the reciprocals of all rational numbers greater than (or equal to) 2. See Figure 6.7.

Wrapping all this together yields the following definition.

Definition 6.14. Let **a** be a positive real number and let L be a representative of **a**. Then $\mathbf{a}^{-1}$ is represented by

$$\{x \in \mathbb{Q}: x \leq 0\} \cup \{x^{-1}: x \in \mathbb{Q},\ x > 0, x \notin \mathbf{a}\}.$$

If **a** is negative, then $\mathbf{a}^{-1} = -[(-\mathbf{a})^{-1}]$.

It takes a lot of effort and checking of cases to show that, indeed, $\mathbf{a} \cdot \mathbf{a}^{-1} = 1$ for any nonzero real number **a**.

Combining all the properties we have considered for $+$, $\cdot$, and $\leq$, we have the following result.

Proposition 6.15. *The real number system* $(\mathbb{R}, +, \cdot, \leq)$ *is an ordered field.*

Both the rationals and the reals are ordered fields, but there's a crucial difference: the reals form a *complete* ordered field, and that's the subject of the next section.

6.6 Completeness

The rational numbers have "holes." If we think of $\mathbb{Q}$ as a number line there is a "gap" at $\sqrt{2}$. There are rational numbers x with $x^2 < 2$ and rational numbers x with $x^2 > 2$ but there's no rational number x with $x^2 = 2$. The real numbers "plug" those "holes" and form a continuous number line; we call the reals a *continuum*. We need to make this precise starting with the notion of an *upper bound*.

Let X be a nonempty set of real numbers. A real number **b** is an *upper bound* for X provided all elements in X are less than or equal to **b**.

For example, let X be the set of negative real numbers. The real number **1** is an upper bound for X because for every $\mathbf{x} \in X$ we know that $\mathbf{x} \leq \mathbf{1}$. Of course, **1** is not the only upper bound for X. The real number **2** is also an upper bound. And, naturally, **0** is an upper bound for X because if $\mathbf{x} \in X$, then $\mathbf{x} \leq \mathbf{0}$.

Zero is an interesting upper bound for the set of negative real numbers, X. While X has many upper bounds, among all the upper bounds **0** is the *least* upper bound. These ideas are important and we emphasize them with a formal definition.

Definition 6.16. Let X be a nonempty set of real numbers. An *upper bound* for X is a real number **b** such that for all $\mathbf{x} \in X$ we have $\mathbf{x} \leq \mathbf{b}$.

Furthermore, we say **b** is a *least upper bound* for X if both of the following hold.

- **b** is an upper bound for X,
- If **b'** is also an upper bound for X, then $\mathbf{b} \leq \mathbf{b'}$.

The least upper bound of X is denoted lub X.

Note that a set cannot have two different least upper bounds; see Exercise 6.18.

Let's consider a few more examples.

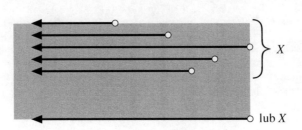

Figure 6.8: In the top portion of this figure we show a collection of left rays that represent the real numbers in a set X. The union of all these representatives rays gives a left ray for the least upper bound of X.
Note that the set X need not be finite. The union of infinitely many left rays is a left ray (provided the union is not all of $\mathbb{Q}$).

- Let X be the *closed* interval $[-1, 1]$. That is, $X = \{\mathbf{x} \in \mathbb{R} \colon -1 \le \mathbf{x} \le 1\}$. Then 1 is the least upper bound for X.

- Let X be the *open* interval $(-1, 1)$. That is, $X = \{\mathbf{x} \in \mathbb{R} \colon -1 < \mathbf{x} < 1\}$. As in the previous example, lub $X = 1$.

- Let X be the set of positive real numbers. Then X does not have any upper bounds.

- Let $X = \{\mathbf{x} \in \mathbb{R} \colon \mathbf{x}^2 < 2\}$. Then lub $X = \sqrt{2}$.

Proposition 6.17 (Completeness property). *Let X be a nonempty subset of $\mathbb{R}$. If X has an upper bound, then X has a least upper bound.*

Let's see why this works.

Let X be a nonempty set of real numbers. Each real number $\mathbf{x}$ in X is represented by left ray $L_\mathbf{x}$. See the top portion of Figure 6.8.

Because X is a bounded above, there is some real number $\mathbf{b}$ that is larger than all the elements in X. This implies that the rays $L_\mathbf{x}$ do not cover all rational numbers. In other words, when we form the union of all the left rays $L_\mathbf{x}$ the result is a left ray, R. This is illustrated in the lower portion of Figure 6.8; the shading can be thought of as the rays above casting a shadow directly below to form R.

It now follows that $\mathbf{p} = [\![R]\!]$ is a real number that is greater than or equal to all members of X; that is, $\mathbf{p}$ is an upper bound for X. Furthermore, any real number $\mathbf{y}$ smaller than $\mathbf{p}$ cannot be an upper bound for X because $\mathbf{y}$'s ray would fail to include all the rays for X.

Therefore $\mathbf{p}$ is the least upper bound for X.

Combining Propositions 6.15 and 6.17 we have the following key result.

Theorem 6.18. *The real number system* $(\mathbb{R}, +, \cdot, \leq)$ *is a complete ordered field.*

The details of being a complete ordered field are summarized in the box on the next page.

6.7 Assimilation

Let t be a rational number. Real numbers represented by rays of the form $\{x \in \mathbb{Q} : x < t\}$ yield an exact replica of the rational numbers, $\mathbb{Q}$. Now that we have created the real numbers, we can "forget" the original rational numbers and, instead, simply see rational numbers as a proper subset of $\mathbb{R}$.

Recap

In this chapter we defined real numbers by means of equivalence classes of left rays, also known as Dedekind cuts. We defined addition, multiplication, and less-than-or-equal for this formulation of real numbers and asserted that $(\mathbb{R}, +, \cdot, \leq)$ form a complete ordered field.

Exercises

6.1 The real number **5** is an equivalence class of left rays. What are those rays?

6.2 Let L_1 and L_2 be left rays. Show that $L = \{x + y : x \in L_1 \text{ and } y \in L_2\}$ is also a left ray.

6.3 In this exercise we illustrate that addition of real numbers does not depend on the choice of representatives. Define the following left rays:

$$A_1 = \{x \in \mathbb{Q} : x < 1\}, \quad A_2 = \{x \in \mathbb{Q} : x < 2\},$$
$$B_1 = \{x \in \mathbb{Q} : x \leq 1\}, \quad B_2 = \{x \in \mathbb{Q} : x \leq 2\}.$$

There are four ways to calculate **1** + **2**:

> ### $(\mathbb{R}, +, \cdot, \leq)$ is a complete ordered field
>
> The real number systems has the following properties.
>
> - Addition and multiplication are commutative: For real numbers **x** and **y**, we have $\mathbf{x} + \mathbf{y} = \mathbf{y} + \mathbf{x}$ and $\mathbf{x} \cdot \mathbf{y} = \mathbf{y} \cdot \mathbf{x}$.
>
> - Addition and multiplication are associative: For real numbers **x**, **y**, and **z**, we have $\mathbf{x} + (\mathbf{y} + \mathbf{z}) = (\mathbf{x} + \mathbf{y}) + \mathbf{z}$ and $\mathbf{x} \cdot (\mathbf{y} \cdot \mathbf{z}) = (\mathbf{x} \cdot \mathbf{y}) \cdot \mathbf{z}$.
>
> - Addition and multiplication have identity elements, **0** and **1** respectively with $\mathbf{0} \neq \mathbf{1}$. That is, for any real number **x** we have $\mathbf{0} + \mathbf{x} = \mathbf{x}$ and $\mathbf{1} \cdot \mathbf{x} = \mathbf{x}$.
>
> - All real numbers have additive inverses, and all nonzero real numbers have multiplicative inverses. That is, for every real number **x** there is a real number $-\mathbf{x}$ such that $\mathbf{x} + (-\mathbf{x}) = \mathbf{0}$ and, if $\mathbf{x} \neq \mathbf{0}$, there is a real number $\mathbf{x}^{-1}$ such that $\mathbf{x} \cdot \mathbf{x}^{-1} = \mathbf{1}$.
>
> - The distributive property holds: For all real numbers **x**, **y**, and **z** we have $\mathbf{x} \cdot (\mathbf{y} + \mathbf{z}) = \mathbf{x} \cdot \mathbf{y} + \mathbf{x} \cdot \mathbf{z}$.
>
> - The $\leq$ relation satisfies the reflexive, antisymmetric, transitive, and trichotomy properties. That is, for all real numbers **x**, **y**, and **z** the following hold: (a) $\mathbf{x} \leq \mathbf{x}$, (b) if $\mathbf{x} \leq \mathbf{y}$ and $\mathbf{y} \leq \mathbf{x}$ then $\mathbf{x} = \mathbf{y}$, (c) if $\mathbf{x} \leq \mathbf{y}$ and $\mathbf{y} \leq \mathbf{z}$ then $\mathbf{x} \leq \mathbf{z}$, and (d) exactly one of the following is true: $\mathbf{x} < \mathbf{y}$, $\mathbf{x} = \mathbf{y}$, or $\mathbf{x} > \mathbf{y}$.
>
> - For positive **x** and **y** both $\mathbf{x} + \mathbf{y}$ and $\mathbf{x} \cdot \mathbf{y}$ are positive.
>
> - If X is a nonempty set of real numbers that has an upper bound, then X has a least upper bound.

- $\{x + y : x \in A_1 \text{ and } y \in B_1\}$,
- $\{x + y : x \in A_1 \text{ and } y \in B_2\}$,
- $\{x + y : x \in A_2 \text{ and } y \in B_1\}$,
- $\{x + y : x \in A_2 \text{ and } y \in B_2\}$.

These four sets are all left rays. What are they? Are they equivalent? What real number(s) do they represent?

6.4 The conversion from decimals to left rays illustrated in the boxed comment on page 87 fails for negative real numbers. Explain what the problem is and propose a solution.

6.5 Illustrate how the real number π can be expressed as a left ray by using its decimal expansion.

6.6 Let L be an open left ray. Sometimes the flip-and-swap operation yields a closed left ray, and sometimes an open left ray. Given an example of each.

6.7 We know that **0** is its own additive inverse because $\mathbf{0} + \mathbf{0} = \mathbf{0}$. Explain, using the flip-and-swap operation, why the additive inverse of **0** is **0**.

6.8 Show that $\mathbf{1}^{-1} = \mathbf{1}$ by using Definition 6.14.

6.9 These two equivalent left rays both represent the real number **2**:

$$L = \{x \in \mathbb{Q} : x < 2\} \quad \text{and} \quad L' = \{x \in \mathbb{Q} : x \leq 2\}.$$

Applying Definition 6.14 to these two rays gives these results:

$$\{x \in \mathbb{Q} : x \leq 0\} \cup \{x^{-1} : x \in \mathbb{Q}, x > 0, x \notin L\}$$

and

$$\{x \in \mathbb{Q} : x \leq 0\} \cup \{x^{-1} : x \in \mathbb{Q}, x > 0, x \notin L'\}.$$

Simplify these two results and show that they are equivalent left rays.

6.10 Let x be an integer. Show that if x is even, then x^2 is also even, but if x is odd, then x^2 is also odd.

6.11 In this exercise we outline an alternative approach to showing that there is no rational number x with $x^2 = 2$. As in the proof of Proposition 6.1, suppose (for contradiction) that $x = b/a$ where a and b are positive integers. In this argument we may assume that a and b are as small as possible.

Fill in the details for these remaining steps:

- Integers a and b have no common factor (common divisor greater than 1). If they did, what could we do?
- From $(b/a)^2 = 2$ we have $b^2 = 2a^2$. Explain why b is even.
- Explain why a is even.
- What's the contradiction?

6.12 Show that $\log_2 3$ is irrational. That is, show that there is no rational number x such that $2^x = 3$.

6.13 Use the fact that a real number is positive exactly when its left rays contain positive rational numbers to show that if $\mathbf{x}$ and $\mathbf{y}$ are positive, then so is $\mathbf{x} + \mathbf{y}$.

6.14 Let $L_2 = \{x \in \mathbb{Q}: x < 2\}$ and $L_3 = \{x \in \mathbb{Q}: x < 3\}$.

What are the elements of the set $\{x \cdot y: x \in L_2 \text{ and } y \in L_3\}$?

6.15 What happens when we multiply a positive real number **x** by **0** using Definition 6.11? Examine two cases: a) use an open left ray for **0**; b) use a closed left ray for **0**.

6.16 Let $X = \{-\frac{1}{2}, -\frac{1}{3}, -\frac{1}{4}, -\frac{1}{5}, \ldots\}$.

a) Describe the left rays $L_\mathbf{x}$ for the real numbers **x** in L.

b) Describe the union of these left rays,
$$R = \bigcup_{\mathbf{x} \in L} L_\mathbf{x}.$$
Is R an open or a closed left ray?

c) What real number is represented by R? Explain why this number is the least upper bound of X.

6.17 Consider the following set:
$$L = \{a \in \mathbb{Q}: a < 0 \text{ and } a^2 \geq 2\}.$$
Verify that L is a left ray and describe the real number it represents.

6.18 Show that if a set of real numbers has a least upper bound, that least upper bound is unique. (There can't be two different least upper bounds for the same set.)

6.19 Show that the rational number system does not satisfy the completeness property.

Addendum: Density of Rational Squares

Our demonstration that there is a real number **a** such that $\mathbf{a}^2 = \mathbf{2}$ relied on the following result.

Proposition 6.19. *Let a and b be rational numbers with $0 < a < b$. Then there is a positive rational number s with $a < s^2 < b$.*

Proof. Since a is a positive rational number, express it as $a = p/q$ where p and q are positive integers.

Let n be a (large) positive integer; we specify it more precisely later in this argument.

Then the following is true:
$$a = \frac{p}{q} = \frac{pq}{q^2} = \frac{n^2 pq}{n^2 q^2}.$$

The denominator of the last fraction, n^2q^2, is a perfect square.

We specify the integer n later. For now, just think of n as being very large.

The numerator n^2pq must lie between two perfect squares; that is, let x be a positive integer with
$$x^2 \leq n^2pq < (x+1)^2. \qquad (*)$$
Because $x^2 \leq n^2pq$ and $n^2pq \leq n^2(pq)^2$ we have $x^2 \leq (npq)^2$, which implies $x \leq npq$. We have:

$$\begin{aligned}(x+1)^2 &= x^2 + 2x + 1 \\ &\leq x^2 + 3x &&\text{because } x \geq 1 \\ &\leq n^2pq + 3x &&\text{by } (*) \\ &\leq n^2pq + 3npq &&\text{because } x \leq npq.\end{aligned} \qquad (**)$$

Let $s = (x+1)/(nq)$. Therefore

$$\begin{aligned}a &= \frac{n^2pq}{n^2q^2} \\ &< \frac{(x+1)^2}{n^2q^2} &&\leftarrow \text{this is } s^2 \\ &\leq \frac{n^2pq + 3npq}{n^2q^2} &&\text{by the calculations in } (**) \\ &= \frac{n^2pq}{n^2q^2} + \frac{3npq}{n^2q^2} \\ &= a + \frac{3npq}{n^2q^2} \\ &= a + \frac{3p}{nq}.\end{aligned}$$

We have shown
$$a < s^2 < a + \frac{3p}{nq}.$$

Now we specify a value for n. Since $b > a$, we know that $b - a$ is a positive rational number; choose a value for n so that
$$\frac{3p}{nq} < b - a.$$

Therefore
$$a < s^2 < a + \frac{3p}{nq} < a + (b-a) = b$$

and the proof is complete. □

Exercise 5.12 asserts that the rational numbers are dense; that is, given two different rational numbers, one can always find a third rational number between them. Proposition 6.19 is another example of a density result: between two distinct positive rational numbers, one can always find a rational perfect square.

Chapter 7

ℝ: Real Numbers II, Cauchy Sequences

Chapter 6 created the real numbers from the rational numbers by way of special types of subsets of $\mathbb{Q}$ called Dedekind cuts.

In this chapter we take an altogether different approach to defining real numbers. We don't actually need a second approach, but it provides an interesting alternative. And it raises the question: Which real number system is the real one (pun intended)?

Because we already gave an in-depth approach to real numbers in the previous chapter, this chapter focuses on main ideas and omits many of the details necessary to prove the correctness of our assertions.

7.1 Cauchy Sequences

The key ingredients in this construction of the real number system are (infinite) *sequences*[1] of rational numbers. Here is an example of such a sequence:

$$\left(1, \tfrac{1}{2}, \tfrac{1}{3}, \tfrac{1}{4}, \tfrac{1}{5}, \ldots\right).$$

In this example there is an implied pattern to the terms in the sequence, but that is not necessary. Any infinitely long list of rational numbers is permitted:

$$\left(\tfrac{2}{3}, 0, \tfrac{18}{5}, -\tfrac{1}{4}, \tfrac{22}{7}, 17, 17, \tfrac{5}{8}, \ldots\right).$$

It is useful to have some notation for sequences. We use $\hat{a}$ to stand for a sequence whose elements are $a_1, a_2, \ldots$; that is,

$$\hat{a} = (a_1, a_2, a_3, \ldots).$$

---
[1] Here is a technical definition: Let $\mathbb{Z}^+$ denote the set of positive integers. A *sequence* of rational numbers is a function $f: \mathbb{Z}^+ \to \mathbb{Q}$. The sequence is $(f(1), f(2), f(3), \ldots)$. In this chapter, all sequences are infinitely long.

Likewise, $\hat{b}$ is the sequence whose elements are b_1, b_2, and so on.

A *Cauchy* sequence is a particular type of sequence named in honor of the nineteenth-century French mathematician Augustin-Louis Cauchy. Roughly speaking, in a Cauchy sequence the difference between the terms gets smaller as we proceed further and further into the sequence. Let's make this precise.

What is a small difference? Perhaps our idea of small is that the terms are within $1/100$ of each other. If a sequence is Cauchy, then from some point on, the difference between any two terms – not just consecutive terms – is always less than $1/100$. Now maybe $1/100$ isn't all that small. Perhaps 10^{-9} is our idea of "small." In a Cauchy sequence from some point on the difference between any two terms is less than 10^{-9}. What about a difference of 10^{-100}? That's certainly small and, indeed, in a Cauchy sequence, from some point on, the difference between any two terms is less than 10^{-100}.

For example, consider the first sequence we presented:

$$\hat{a} = \left(1, \tfrac{1}{2}, \tfrac{1}{3}, \tfrac{1}{4}, \tfrac{1}{5}, \ldots\right).$$

Is there a point in $\hat{a}$ after which the differences are less than $1/100$? The nth term in $\hat{a}$ is $a_n = 1/n$. The difference between the nth and mth terms is $|a_n - a_m| = \left|\tfrac{1}{n} - \tfrac{1}{m}\right|$.

- If $n > m$, then $\tfrac{1}{m} > \tfrac{1}{n}$ and so $\left|\tfrac{1}{n} - \tfrac{1}{m}\right| = \tfrac{1}{m} - \tfrac{1}{n} < \tfrac{1}{m}$.

- If $n < m$, then $\tfrac{1}{m} < \tfrac{1}{n}$ and so $\left|\tfrac{1}{n} - \tfrac{1}{m}\right| = \tfrac{1}{n} - \tfrac{1}{m} < \tfrac{1}{n}$.

- And, of course, if $n = m$ then $\left|\tfrac{1}{n} - \tfrac{1}{m}\right| = 0$.

From this we see that if both n and m are greater than 100, then $|a_n - a_m| < 1/100$. That is, once we reach term 101, all differences between all terms are less than $1/100$.

We see that if n and m are greater than 10^{100} then $|a_n - a_m| < 10^{-100}$. Therefore, after we skip the mere first googol terms, the difference between any two terms is less than 10^{-100}.

A sequence is a Cauchy sequence if, given any notion of "small," from some point, on the difference between any pair of terms is small. Here is the mathematical definition.

Definition 7.1 (Cauchy sequence). Let $\hat{a} = (a_1, a_2, a_3, \ldots)$ be a sequence of rational numbers. We say that $\hat{a}$ is a *Cauchy* sequence if for every positive rational number s there is a positive integer N such that for all $n, m > N$ we have $|a_n - a_m| < s$.

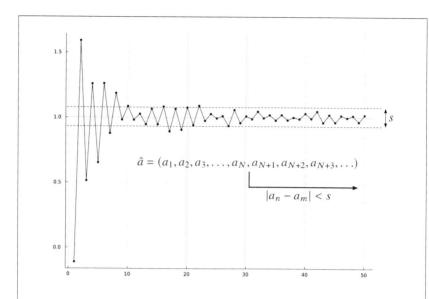

Figure 7.1: This is an illustration of a Cauchy sequence, $\hat{a}$. Given any positive rational number s (a "small" number), there is a point N after which all differences between terms of $\hat{a}$ are less than s.
This can be visualized by a horizontal strip. For any given thickness $s > 0$, from some point on, all the values of the sequence lie inside a strip of thickness s as shown in the diagram.

In this definition, the number s is an idea of "small." Given s, if we ignore the first N terms, all differences between terms are small. To be Cauchy means that, no matter how small this positive number s might be, from some point on in the sequence, the differences are all less than s. The number N typically depends on s. This is illustrated in Figure 7.1.

It is important to emphasize that in Definition 7.1 we require that, from some point on, the differences between *all* pairs of entries is small. Merely showing that the difference between *consecutive* terms is small is insufficient. This issue is explored in Exercise 7.8.

Decimal Numbers as Cauchy Sequences

There is a simple way decimal numbers can be interpreted as Cauchy sequences. For example, in base ten the number π is

$$\pi = 3.1415926535897932\ldots$$

> **Eventually**
>
> A word that helps to clarify the concept of a Cauchy sequence is *eventually*. The beginning of a Cauchy sequence is completely irrelevant. The first million terms might be very far apart from each other. It matters only that *eventually* the differences between all terms are small.
>
> Pick a small positive number ... any small positive number, say 0.00001. Then *eventually* the differences between any two terms in a Cauchy sequence are less than 0.00001. Pick an even smaller number, say $s = 10^{-100}$. We might need to ignore the first trillion terms in the sequence, but *eventually* the differences between any two members of the sequence are less than s. See Exercise 7.4, which underscores that the first term – or the first billion terms – of a Cauchy sequence are irrelevant. All that matters is how it behaves *eventually*.

With each additional digit, the decimal approximation comes closer and closer to the true value of π. We can write these approximations as a sequence of rational numbers:

$$\hat{p} = \left(3, \frac{31}{10}, \frac{314}{100}, \frac{3141}{1000}, \frac{31415}{10000}, \frac{314159}{100000}, \ldots\right).$$

This is more legible using decimals:

$$\hat{p} = (3, 3.1, 3.14, 3.141, 3.1415, 3.14159, \ldots).$$

The difference between, say, 3.14 and any subsequent term involves the third and subsequent digits after the decimal point. That is, the difference between any two terms from 3.14 and onward is of the form 0.00xxxxx..., and so that difference is less that $0.01 = 1/100$.

Likewise, the difference between any two terms from 3.141 and onward is less than $1/1,000$. In this way, if we want to ensure that all pairwise differences are less than some positive rational number s, we simply need to find a power of ten less than s, that is, $10^{-k} < s$. Then from the kth term onward in $\hat{p}$, all pairwise differences are less than s. Hence, $\hat{p}$ is a Cauchy sequence.

There's nothing special about $\hat{p}$. Any decimal expression gives rise to a Cauchy sequence by adding digits one by one.

This foreshadows our next step. In a sense, the Cauchy sequence $\hat{p}$ "is" the number π. However, there's a difficulty. Instead of taking additional digits one at a time, we could take them two at a time, like this:

$$\hat{q} = (3.1, 3.141, 3.14159, 3.1415926, 3.141592653, \ldots).$$

As sequences, $\hat{p} \neq \hat{q}$, yet we would like to think they both represent π.

Here's another example. Consider these two Cauchy sequences built from two different decimal representations of the number 1:

$$\hat{a} = (1, 1, 1, 1, 1, \ldots),$$
$$\hat{b} = (0, 0.9, 0.99, 0.999, 0.9999, \ldots).$$

Again, $\hat{a} \neq \hat{b}$, but we like to think both of them "are" the number 1.

The solution is to define an equivalence relation so that the two sequences derived from π are equivalent to each other, just as the two sequences derived from the decimal representations of 1 are equivalent: $\hat{p} \equiv \hat{q}$ and $\hat{a} \equiv \hat{b}$.

7.2 Equivalent Cauchy Sequences

Just as $1.0000\ldots$ and $0.9999\ldots$ are both representations of the number 1, the Cauchy sequences we derive from them provide two ways to represent the number 1:

$$(1, 1, 1, \ldots) \quad \text{and} \quad (0, 0.9, 0.99, 0.999, 0.9999, \ldots).$$

While these two sequences are not the same, we want to consider them to be equivalent. To this end, we define an equivalence relation $\equiv$ for Cauchy sequences.

Line up the terms of the two sequences like this:

$$\begin{array}{ccccccc} 1 & 1 & 1 & 1 & 1 & & \ldots \\ \updownarrow & \updownarrow & \updownarrow & \updownarrow & \updownarrow & & \\ 0 & 0.9 & 0.99 & 0.999 & 0.9999 & & \ldots \end{array}$$

Notice that the difference between corresponding terms becomes smaller and smaller. Can we make the difference be less than $s = 10^{-100}$? Yes! We just need to go a few steps past the hundredth term and, from that point on, the differences between corresponding terms are all less than s.

Definition 7.2 (Equivalent Cauchy sequences). Let $\hat{a}$ and $\hat{b}$ be Cauchy sequences. We say these sequences are *equivalent*, and we write $\hat{a} \equiv \hat{b}$, if for every positive rational number s, there is a positive integer N such that for all $n > N$ we have $|a_n - b_n| < s$.

Figure 7.2 illustrates this definition. Given any "small" positive number s we choose, after some point the corresponding terms of the two sequences are within s of each other.

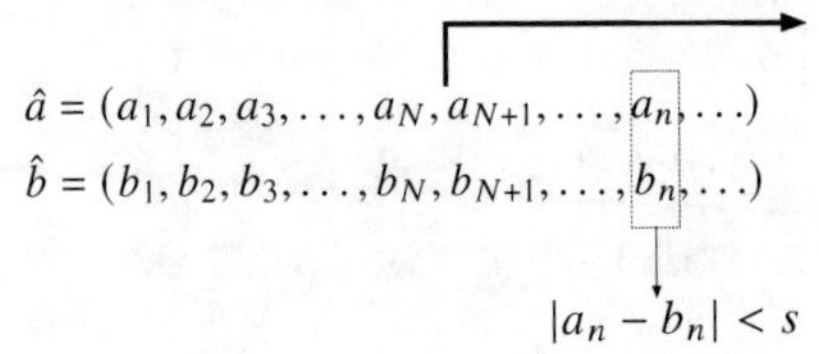

$$\hat{a} = (a_1, a_2, a_3, \ldots, a_N, a_{N+1}, \ldots, a_n, \ldots)$$
$$\hat{b} = (b_1, b_2, b_3, \ldots, b_N, b_{N+1}, \ldots, b_n, \ldots)$$
$$|a_n - b_n| < s$$

Figure 7.2: This is an illustration of equivalent Cauchy sequences, $\hat{a} \equiv \hat{b}$. Given any positive rational number s (a "small" number), there is a point N after which the differences between corresponding terms is always less than s. Informally, we say that *eventually* all differences are small.

To justify calling this relation *equivalence* of Cauchy sequences, we should verify that $\equiv$ is, indeed, an equivalence relations by checking it is reflexive, symmetric, and transitive.

The verification of the reflexive and symmetric properties is not difficult. For the transitive property, we use the following result.

Proposition 7.3 (Triangle inequality). *Let x, y be rational numbers. Then $|x + y| \le |x| + |y|$.*

The verification of this proposition is left to you; see Exercise 7.1. Note that the triangle inequality also holds for real numbers, but at this point those have not been defined.

We now verify that $\equiv$ is an equivalence relation.

- *Reflexive*: Let $\hat{a}$ be a Cauchy sequence. Clearly $\hat{a} \equiv \hat{a}$ because the term-by-term differences are all zero, and zero is less than any positive rational number s.

- *Symmetric*: Let $\hat{a}$ and $\hat{b}$ be Cauchy sequences with $\hat{a} \equiv \hat{b}$. Notice that $|a_n - b_n| = |b_n - a_n|$ so the term-by-term difference are unchanged. It follows that $\hat{b} \equiv \hat{a}$.

- *Transitive*: This is more complicated. Suppose $\hat{a} \equiv \hat{b}$ and $\hat{b} \equiv \hat{c}$. We want to demonstrate that $\hat{a} \equiv \hat{c}$.

  Let s be a positive rational number. We need to find a number N so that for all $n > N$ we have $|a_n - c_n| < s$.

Because $\hat{a} \equiv \hat{b}$, there is a positive integer N_1 after which we have $|a_n - b_n| < s/2$. Likewise, because $\hat{b} \equiv \hat{c}$ there is an N_2 so that for all $n > N_2$ we have $|b_n - c_n| < s/2$.

Let N be the larger of N_1 or N_2. Then for $n > N$ we have

$$\begin{aligned}|a_n - c_n| &= |(a_n - b_n) + (b_n - c_n)| \\ &\leq |a_n - b_n| + |b_n - c_n| \quad \text{by the triangle inequality} \\ &< \frac{s}{2} + \frac{s}{2} = s.\end{aligned}$$

Therefore $\equiv$ is, indeed, an equivalence relation for Cauchy sequences. The next step is to define real numbers. Just as we defined natural numbers, integers, and rational numbers as equivalence classes, we do the same here.

Definition 7.4. A *real number* is an equivalence class of Cauchy sequences. The set of all real numbers is denoted $\mathbb{R}$.

This definition is, of course, utterly different from Definition 6.4, which asserted that Dedekind cuts (equivalence classes of left rays of rational numbers) are real numbers. We deal with this seeming inconsistency in the next chapter. For now, let's see how to add, multiply, and compare real numbers with this definition.

7.3 Addition and Multiplication

This chapter's definition of real number is more complicated than the definition from the previous chapter. The good news is that the definitions of addition and multiplication are simpler.

We begin by defining addition and multiplication of Cauchy sequences. Let $\hat{a}$ and $\hat{b}$ be Cauchy sequences. We define their sum and product simply by forming their term-by-term sum and product; by this we mean

$$\hat{a} + \hat{b} = (a_1 + b_1, a_2 + b_2, a_3 + b_3, \ldots),$$
$$\hat{a} \cdot \hat{b} = (a_1 b_1, a_2 b_2, a_3 b_3, \ldots).$$

The sum and product of Cauchy sequences are, themselves, Cauchy sequences.

Proposition 7.5. *Let $\hat{a}$ and $\hat{b}$ be Cauchy sequences. Then $\hat{a} + \hat{b}$ and $\hat{a} \cdot \hat{b}$ are also Cauchy sequences.*

To show that $\hat{c} = \hat{a} + \hat{b}$ is a Cauchy sequence, let s be a positive rational number. Because $\hat{a}$ is a Cauchy sequence, there is a positive integer N_1 such that for $n, m > N_1$ we have $|a_n - a_m| < s/2$. Likewise, there is a positive integer N_2 so that for all $n, m > N_2$ we have $|b_n - b_m| < s/2$.

Let N be the larger of N_1 or N_2. Then for $n, m > N$ we have

$$\begin{aligned} |c_n - c_m| &= |(a_n + b_n) - (a_m + b_m)| \\ &= |(a_n - a_m) + (b_n - b_m)| \\ &\leq |a_n - a_m| + |b_n - b_m| \quad \text{by the triangle inequality} \\ &< \frac{s}{2} + \frac{s}{2} = s. \end{aligned}$$

The verification that $\hat{a} \cdot \hat{b}$ is a Cauchy sequence is more complicated. We omit the proof; see Exercise 7.7.

The definition of addition and multiplication of real numbers rests on addition and multiplication of Cauchy sequences.

Definition 7.6. Let $\mathbf{a} = [\![\hat{a}]\!]$ and $\mathbf{b} = [\![\hat{b}]\!]$ be real numbers. We define their sum and product by

$$\mathbf{a} + \mathbf{b} = [\![\hat{a} + \hat{b}]\!] \quad \text{and} \quad \mathbf{a} \cdot \mathbf{b} = [\![\hat{a} \cdot \hat{b}]\!].$$

As usual, to check that this is a proper definition we need to demonstrate that the results of $\mathbf{a} + \mathbf{b}$ and of $\mathbf{a} \cdot \mathbf{b}$ do not depend on the choice of representatives. That is, if $\hat{a}, \hat{a}' \in \mathbf{a}$ and $\hat{b}, \hat{b}' \in \mathbf{b}$ then

$$[\![\hat{a} + \hat{b}]\!] = [\![\hat{a}' + \hat{b}']\!] \quad \text{and} \quad [\![\hat{a} \cdot \hat{b}]\!] = [\![\hat{a}' \cdot \hat{b}']\!].$$

For addition, let $\hat{c} = \hat{a} + \hat{b}$ and $\hat{c}' = \hat{a}' + \hat{b}'$. We need to show that $\hat{c} \equiv \hat{c}'$.

This means, given a positive rational number s, we need to find an integer N so that for $n > N$ we have $|c_n - c_n'| < s$.

Because $\hat{a} \equiv \hat{a}'$, there is an integer N_1 so that for $n > N_1$ we have $|a_n - a_n'| < s/2$. Likewise, there is an integer N_2 so that for $n > N_2$ we have $|b_n - b_n'| < n/2$. Therefore

$$\begin{aligned} |c_n - c_n'| &= |(a_n + b_n) - (a_n' + b_n')| \\ &= |(a_n - a_n') + (b_n - b_n')| \\ &\leq |a_n - a_n'| + |b_n - b_n'| \\ &< \frac{s}{2} + \frac{s}{2} = s. \end{aligned}$$

We omit the verification for multiplication. It has a similar flavor, but the algebra is more complicated.

Having defined addition and multiplication of real numbers, the next step is to verify that $(\mathbb{R}, +, \cdot)$ is a field (see Definition 5.4). We present highlights.

To begin, it's not hard to see that addition and multiplication are commutative and associative.

Addition has an identity element: $\mathbf{0} = [\![(0, 0, 0, 0, \ldots)]\!]$. It is not difficult to check that for any real number $\mathbf{x} + \mathbf{0} = \mathbf{x}$.

Every real number has an additive inverse. For a given real number $\mathbf{x} = [\![\hat{x}]\!]$ with $\hat{x} = (x_1, x_2, x_3, \ldots)$, let $-\hat{x} = (-x_1, -x_2, -x_3, \ldots)$. It follows that $[\![-\hat{x}]\!]$ is the additive inverse of $\mathbf{x}$, namely $-\mathbf{x}$.

Multiplication has an identity element: $\mathbf{1} = [\![(1, 1, 1, 1, \ldots)]\!]$. Checking that $\mathbf{x} \cdot \mathbf{1} = \mathbf{x}$ is not difficult.

Multiplicative inverses are tricky. Let $\mathbf{x}$ be a real number with $\mathbf{x} \neq \mathbf{0}$. Suppose $\mathbf{x} = [\![(x_1, x_2, x_3, \ldots)]\!]$. It is tempting (but incorrect) to propose that $[\![(1/x_1, 1/x_2, 1/x_3, \ldots)]\!]$ be the multiplicative inverse of $\mathbf{x}$. The difficulty is that we cannot guarantee that none of the x_ns are zero. Here are the key ideas:

- Since $[\![\hat{x}]\!] \neq \mathbf{0}$, we show (see Exercise 7.11) that $\hat{x}$ has at most finitely many entries equal to zero.

- Let N be the position of the last zero entry in $\hat{x}$.

- Let $\hat{x}'$ be the sequence formed by dropping the first N entries in $\hat{x}$. By Exercise 7.4, $\hat{x}' \equiv \hat{x}$.

- Let $\hat{y} = (1/x_{N+1}, 1/x_{N+2}, 1/x_{N+3}, \ldots)$. Verify that $\hat{y}$ is a Cauchy sequence.

- Conclude that $[\![\hat{y}]\!]$ is a multiplicative inverse of $\mathbf{x}$.

Finally, verification of the distributive property, $\mathbf{x} \cdot (\mathbf{y} + \mathbf{z}) = \mathbf{x} \cdot \mathbf{y} + \mathbf{x} \cdot \mathbf{z}$, is not difficult.

7.4 Ordering Real Numbers

Having defined addition and multiplication, we next consider the less-than-or-equal relation, $\leq$. We begin with some *incorrect* definitions to explain why a modestly more elaborate definition is necessary.

First failed attempt: *Let $\mathbf{a}$ and $\mathbf{b}$ be real numbers and let $\hat{a} \in \mathbf{a}$ and $\hat{b} \in \mathbf{b}$. We say that $\mathbf{a} \leq \mathbf{b}$ if for all n we have $a_n \leq b_n$.* This looks clean and is similar to how we defined addition and multiplication. Here's the problem. Consider these two real numbers and their representatives:

$$\mathbf{a} = [\![(2, 1, 1, 1, 1, \ldots)]\!] \quad \text{and} \quad \mathbf{b} = [\![(1, 2, 2, 2, 2, \ldots)]\!].$$

According to the proposed definition, we have neither $\mathbf{a} \leq \mathbf{b}$ nor $\mathbf{b} \leq \mathbf{a}$. The problem is that the important part of a Cauchy sequence is in its tail and not in its beginning. Let's try again.

Second failed attempt: *Let* **a** *and* **b** *be real numbers and let* $\hat{a} \in$ **a** *and* $\hat{b} \in$ **b**. *We say that* **a** $\leq$ **b** *provided there is an integer N such that for all* $n > N$ *we have* $a_n \leq b_n$. This is better because it allows us to ignore the initial portions of $\hat{a}$ and $\hat{b}$, and just focus on their tails. However, consider these two real numbers and their representatives:

$$\mathbf{a} = [\![(0,0,0,0,0,\ldots)]\!] \quad \text{and} \quad \mathbf{b} = [\![(1, -\tfrac{1}{2}, \tfrac{1}{3}, -\tfrac{1}{4}, \tfrac{1}{5}, -\tfrac{1}{6}, \ldots)]\!].$$

By this new faulty definition, we again have the problem that neither **a** $\leq$ **b** nor **b** $\leq$ **a**.

Still, this attempt gets us closer. Let's focus on strict inequality; that is, we want to define when **a** < **b**.

Let $\hat{a}$ and $\hat{b}$ be Cauchy sequences representing the real numbers **a** and **b**. For **a** to be "truly" less than **b** we need that "eventually" the elements in $\hat{b}$ be "truly" larger than the corresponding elements in $\hat{a}$. Here's how to make that idea precise.

First, "eventually" means that the condition we develop needs to take hold starting at some index N. And second, "truly" means that the difference between corresponding terms is at least some positive value s. Here is the formal definition.

Definition 7.7. Let **a** and **b** be real numbers represented by Cauchy sequences $\hat{a}$ and $\hat{b}$. We say that **a** *is less* than **b** (and we write **a** < **b**) provided there is a positive rational number s and a positive integer N such that for all indices $i > N$ we have $b_i - a_i \geq s$.

From this, we define **a** $\leq$ **b** to mean that either **a** = **b** or **a** < **b**, and similarly for > and $\geq$. We say that a real number **a** is *positive* if **a** > **0** and *negative* if **a** < **0**.

From this definition we immediately have this:

Proposition 7.8. *Let* **x** *be a real number represented by* $\hat{x}$.

Then **x** *is positive if and only if there is a positive rational number s and a positive integer N such that* $x_i > s$ *for all* $i \geq N$.

Likewise, **x** *is negative if and only if there is a positive rational number s and a positive integer N such that* $x_i < -s$ *for all* $i \geq N$.

For example, consider the following Cauchy sequence:

$$\hat{a} = \left(\frac{-7}{8}, \frac{5}{8}, \frac{-1}{8}, \frac{1}{4}, \frac{1}{16}, \frac{5}{32}, \frac{7}{64}, \frac{17}{128}, \frac{31}{256}, \ldots\right).$$

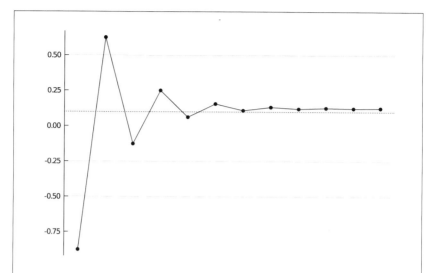

Figure 7.3: This graph shows the successive terms of a Cauchy sequence $\hat{a}$. Although the sequence contains negative values, eventually all terms are greater than $1/10$ as indicated by the horizontal dotted line. Hence $[\![\hat{a}]\!]$ is a positive real number.

The nth term of this sequence is given by the formula $a_n = \frac{1}{8} - \left(\frac{-1}{2}\right)^{k-1}$ but it is best understood by plotting the terms as shown in Figure 7.3. Notice that some terms in the sequence are negative. Nevertheless, from some point onward, all the terms are greater than $1/10$ and therefore $[\![\hat{a}]\!]$ is positive.

On the other hand, consider the sequence

$$\hat{b} = \left(1, \frac{1}{2}, \frac{1}{3}, \frac{1}{4}, \frac{1}{5}, \frac{1}{6}, \ldots\right).$$

All the terms in this sequence are positive, but it is not the representative of a positive real number. If it were, eventually all the terms would be larger than s for some positive rational number s. But we see that when n is large enough $1/n < s$. Indeed, $[\![\hat{b}]\!] = \mathbf{0}$.

7.5 Bisection and our Favorite Real Number

We know there is no rational square root of 2, but we can find good approximations. One way to do this is by the *bisection* method.

116   From Counting to Continuum

To begin, notice that the function $f(x) = x^2$ is increasing for positive values of x. That is, if x and y are positive, and if $x < y$, then $x^2 < y^2$. So if we guess a possible value for $\sqrt{2}$ that is too small, we should increase it; conversely, if we guess a value that's too large, we should decrease it.

Let's begin with two guesses: 1 and 2. Because $1^2 = 1$ that guess is too small. And because $2^2 = 4$, that guess is too large. That means, if we're going to find a square root of 2, we need to look between those values.

A reasonable middle ground is the midpoint between these two numbers, namely $\frac{3}{2}$. Notice that $\left(\frac{3}{2}\right)^2 = \frac{9}{4} > 2$. So $\frac{3}{2}$ is also too big, so we should continue our hunt between 1 (which is too small) and $\frac{3}{2}$ (which is too big).

Next step is to pick a value halfway between 1 and $\frac{3}{2}$, namely $\frac{5}{4}$. We calculate $\left(\frac{5}{4}\right)^2 = \frac{25}{16} < 2$. So $\frac{5}{4}$ is too small.

Now we need to explore between $\frac{5}{4}$ and $\frac{3}{2}$; the value midway between them is $\frac{11}{8}$.

Let's make the bisection method more precise. We begin with two guesses that we name $a = 1$ and $b = 2$. We choose these knowing that $a^2 < 2 < b^2$.

Next we consider the value midway between them, namely $m = (a + b)/2$. We square m and see what happens. There are two possibilities:

- $m^2 < 2$. Because m is bigger than a, we know that m^2 is closer to 2 than a^2 is. So we know the elusive value we seek lies between m and b, and this is a tighter range than we had before. So we redefine a to be this new value m and try again.

- $m^2 > 2$. Now we note that m is less than b, so m^2 is closer to 2 than b^2 is. In this case, the elusive value we seek lies between a and m, and so we redefine b to be this new value m.

This process is illustrated in Figure 7.4.

The bisection method produces three streams of rational numbers: the a-values, the m-values, and the b-values:

$$\hat{a} = \left(1, 1, \tfrac{5}{4}, \tfrac{11}{8}, \tfrac{11}{8}, \tfrac{45}{32}, \tfrac{45}{32}, \tfrac{181}{128}, \tfrac{181}{128}, \tfrac{181}{128}, \ldots\right),$$

$$\hat{m} = \left(\tfrac{3}{2}, \tfrac{5}{4}, \tfrac{11}{8}, \tfrac{23}{16}, \tfrac{45}{32}, \tfrac{91}{64}, \tfrac{181}{128}, \tfrac{363}{256}, \tfrac{725}{512}, \tfrac{1449}{1024}, \ldots\right),$$

$$\hat{b} = \left(2, \tfrac{3}{2}, \tfrac{3}{2}, \tfrac{3}{2}, \tfrac{23}{16}, \tfrac{23}{16}, \tfrac{91}{64}, \tfrac{91}{64}, \tfrac{363}{256}, \tfrac{725}{512}, \ldots\right).$$

Notice that at each step the difference between a_k and b_k shrinks by a factor of 2 because we have cut the search interval exactly in half. Hence we have $b_k - a_k = 1/2^{k-1}$.

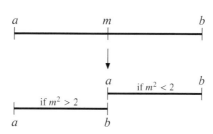

Figure 7.4: Given rational numbers a (with $a^2 < 2$) and b (with $b^2 > 2$), we shrink the search interval by considering its midpoint, m. If $m^2 < 2$, then m becomes the new left end point. If $m^2 > 2$, then m becomes the new right end point.

$a_1 \leftarrow 1$
$b_1 \leftarrow 2$
$m_1 \leftarrow (a_1 + b_1)/2$
for $k = 2$ to ∞ **do**
  if $m_{k-1}^2 > 2$ **then**
    $a_k \leftarrow a_{k-1}$
    $b_k \leftarrow m_{k-1}$
  else
    $a_k \leftarrow m_{k-1}$
    $b_k \leftarrow b_{k-1}$
  $m_k \leftarrow (a_k + b_k)/2$

Figure 7.5: This algorithm produces three Cauchy sequences of rational numbers: $\hat{a}$, $\hat{m}$, and $\hat{b}$. The terms in these sequences give better and better approximate solutions to the equation $x^2 = 2$.

We claim that $\hat{a}$ is a Cauchy sequence. Here's why.

The terms in $\hat{a}$ are confined to smaller and smaller intervals. All terms from a_k onward are between a_k and b_k, and the separation between those values is $1/2^{k-1}$. Hence the difference between any two terms in $\hat{a}$ at position k or beyond is at most $1/2^{k-1}$. Therefore $\hat{a}$ is a Cauchy sequence.

The same analysis shows that $\hat{m}$ and $\hat{b}$ are also Cauchy sequences.

Furthermore, all three are equivalent Cauchy sequences (see Definition 7.2). To show, for example, that $\hat{a} \equiv \hat{b}$, simply recall that the difference between

a_k and b_k is $1/2^{k-1}$. Therefore, from term k onwards, all differences between corresponding terms are less than $1/2^{k-1}$.

Having established that $\hat{a}$, $\hat{m}$, and $\hat{b}$ are equivalent Cauchy sequences, let's consider the real number $\mathbf{x} = [\![\hat{a}]\!] = [\![\hat{b}]\!]$.

We claim that $\mathbf{x}^2 = \mathbf{2}$.

The simplest Cauchy sequence representing $\mathbf{2}$ is $\hat{c} = (2, 2, 2, \ldots)$.

We have two representatives for $\mathbf{x}^2$.

On the one hand, $\mathbf{x}^2$ is represented by $(a_1^2, a_2^2, a_3^2, \ldots)$, which is term-by-term less than $\hat{c}$. Therefore $\mathbf{x}^2 \leq \mathbf{2}$. (See Exercise 7.13.)

On the other hand, $\mathbf{x}^2$ is also represented by $(b_1^2, b_2^2, b_3^2, \ldots)$, which is term by term greater than $\hat{c}$. Therefore $\mathbf{x}^2 \geq \mathbf{2}$.

Since we have both $\mathbf{x}^2 \leq \mathbf{2}$ and $\mathbf{x}^2 \geq \mathbf{2}$, we conclude that $\mathbf{x}^2 = \mathbf{2}$. Therefore $\mathbf{x} = \sqrt{\mathbf{2}}$. And so, once again, we have this:

Proposition 7.9. *The real numbers contain a square root of* $\mathbf{2}$.

7.6 Assimilation and Completeness

We have developed the definitions of real number, the operations of addition and multiplication, and the ordering relation $\leq$. Our final task is to show that the real numbers form a complete ordered field. To that end, we revisit the concepts of upper bound and least upper bound presented earlier in Definition 6.16.

Given a nonempty set of real numbers X, we say that $\mathbf{b}$ is an *upper bound* for X provided every element of X is less than or equal to $\mathbf{b}$. Furthermore, we say that $\mathbf{b}$ is a *least upper bound* for X provided $\mathbf{b}$ is an upper bound and if $\mathbf{b}'$ is any other upper bound for X, then $\mathbf{b} \leq \mathbf{b}'$.

The completeness property states that if X is a nonempty set of real numbers that has an upper bound, then it has a least upper bound. To explain why this is the case, we revisit the bisection method.

First, though, we note that the real numbers contain a copy of the rational numbers. That is, if x is any rational number, then $\hat{x} = (x, x, x, \ldots)$ is a Cauchy sequence and therefore represents a real number $\mathbf{x}$. There is a "copy" of every rational number embedded in $\mathbb{R}$.

In this way, we can compare rational numbers and real numbers because we can think of 1 and $\mathbf{1}$ as the same.

Let X be a nonempty set of real numbers that has an upper bound. We use bisection to show that X has a least upper bound.

To begin, let b_1 be a rational upper bound. In other words, we know that X has some upper bound; let b_1 be any rational number at least as large as that upper bound.

Next, let a_1 be a rational number that is *not* an upper bound. That is, we can take a_1 to be any rational number that is less than some member of X.

So now we have two rational numbers $a_1 < b_1$. All elements of X are less than or equal to b_1 but some elements of X are not less than or equal to a_1.

The next step is to define $m_1 = (a_1 + a_2)/2$. This is a rational number between a_1 and b_1. There are two possibilities:

- m_1 is an upper bound for X. In this case we set $a_2 = a_1$ and $b_2 = m_1$.

- m_1 is not an upper bound for X. In this case, we set $a_2 = m_2$ and $b_2 = b_1$.

In both cases, we have $a_2 < b_2$ where b_2 is an upper bound for X but a_2 is not. Notice that the difference $b_2 - a_2$ is half as large as $b_1 - a_1$.

We now define a_3 and b_3, and so forth, by the same idea. Let $m_k = (a_k + b_k)/2$.

- If m_k is an upper bound for X, let $b_{k+1} = m_k$ and $a_{k+1} = a_k$.

- If m_k is not an upper bound for X, let $a_{k+1} = m_k$ and $b_{k+1} = b_k$.

At each stage the size of the interval between a_k and b_k shrinks by half. Therefore, by the same reasoning we used in Section 7.5, we see that $\hat{a} = (a_1, a_2, a_3, \ldots)$ and $\hat{b} = (b_1, b_2, b_3 \ldots)$ are equivalent Cauchy sequences giving a real number $\mathbf{b} = [\![\hat{a}]\!] = [\![\hat{b}]\!]$.

If $\mathbf{x}$ is any real number in X, then by construction every element of $\hat{b}$ is greater than or equal to $\mathbf{x}$ and so $\mathbf{b}$ is an upper bound for X. Further, there cannot be a smaller upper bound (say, $\mathbf{c}$) for X because otherwise some a_k would be larger than $\mathbf{c}$, implying that there is an element of X that's bigger than $\mathbf{c}$.

Therefore X has a least upper bound.

Theorem 7.10. *The real number system* $(\mathbb{R}, +, \cdot, \leq)$ *is a complete ordered field.*

See the boxed comments on page 100 for the full definition of a complete ordered field.

Although the wording in Theorems 6.18 and 7.10 is exactly the same, these results are different because the first refers to the real number system of Dedekind cuts and the second to the real number system of equivalence classes of Cauchy sequences.

This may leave you wondering: Which definition should we use? Are real numbers Dedekind cuts? Or are they Cauchy sequences? The great answer is: It doesn't matter! And that is the subject of the next chapter.

Recap

In this chapter we presented an entirely different definition of the real numbers. Roughly speaking, we defined Cauchy sequences as infinitely long lists of rational numbers that get closer and closer together to each other. From there we defined what it means for two Cauchy sequences to be equivalent. Real numbers are equivalence classes of Cauchy sequences of rational numbers. We defined addition, multiplication, and less-than-or-equal for real numbers, and explained how nonempty sets of real numbers that have an upper bound must have a least upper bound. Hence, the real numbers constitute a complete ordered field.

Exercises

7.1 Verify the *triangle inequality*: $|x + y| \le |x| + |y|$. You may assume that x and y are rational numbers.

7.2 Show that $|x - y| \le |x| + |y|$.

7.3 Let $\hat{a} = (a_1, a_2, a_3, \ldots)$ be defined by

$$a_k = 1 + \frac{(-1)^k}{k}.$$

That is, $\hat{a} = (0, \frac{3}{2}, \frac{2}{3}, \frac{5}{4}, \frac{4}{5}, \frac{7}{6}, \frac{6}{7}, \ldots)$.
Show that $\hat{a}$ is a Cauchy sequence and that $\hat{a} \equiv (1, 1, 1, 1, \ldots)$.

7.4 Let $\hat{a} = (a_1, a_2, a_3, \ldots)$ be a Cauchy sequence. Show that $\hat{b} = (a_2, a_3, a_4, \ldots)$ is also a Cauchy sequence and that $\hat{a} \equiv \hat{b}$. In other words, if we drop the first term of a Cauchy sequence, the remainder is an equivalent Cauchy sequence.

More radically, show that $\hat{c} = (a_2, a_4, a_6, \ldots)$ is a Cauchy sequence and $\hat{a} \equiv \hat{c}$. In other words, if we delete all the odd-numbered terms of a Cauchy sequence, the terms that remain are an equivalent Cauchy sequence.

7.5 Show that Cauchy sequences are *bounded*, by which we mean this: If $\hat{a}$ is a Cauchy sequence of rational numbers, there is a positive rational number B such that $|a_i| \le B$ for all i.

7.6 Let $\hat{a} = (a_1, a_2, a_3, \ldots)$ be a bounded, increasing sequence of rational numbers. Show that $\hat{a}$ is a Cauchy sequence.

Notes:

- As defined in Exercise 7.5, to say that $\hat{a}$ is *bounded* means there is a rational number b such that all elements of the sequence are less than or equal to b, that is, for all n, $|a_n| \le b$.

- To say that $\hat{a}$ is *increasing* means that $a_1 < a_2 < a_3 < \cdots$.
- Hint: Show that if an increasing sequence $\hat{a}$ were not Cauchy, then it would not be be bounded.

7.7 Let $\hat{a} = (a_1, a_2, a_3, \ldots)$ and $\hat{b} = (b_1, b_2, b_3, \ldots)$ be Cauchy sequences of rational numbers. Show that $(a_1b_1, a_2b_2, a_3b_3, \ldots)$ is also a Cauchy sequence.

Hint: Use the fact that Cauchy sequences are bounded (Exercise 7.5) and this algebraic identity:

$$a_i b_i - a_j b_j = (a_i - a_j)b_i + (b_i - b_j)a_j.$$

7.8 For a positive integer n, the nth *harmonic number* is defined as

$$h_n = 1 + \tfrac{1}{2} + \tfrac{1}{3} + \cdots + \tfrac{1}{n}.$$

Let $\hat{h} = (h_1, h_2, h_3, \ldots)$. Notice that the difference between consecutive terms gets smaller and smaller:

$$|h_n - h_{n+1}| = \frac{1}{n+1}.$$

Is $\hat{h}$ a Cauchy sequence? Justify your answer.

7.9 Let $\hat{a}$ be a Cauchy sequence that contains infinitely many 0s. Show that $\hat{a} \equiv (0, 0, 0, \ldots)$.

7.10 Let $\hat{a}$ be a Cauchy sequence with infinitely many positive values and infinitely many negative values. Show that $\hat{a} \equiv (0, 0, 0, \ldots)$.

7.11 Let **x** be a nonzero real number and let $\hat{x} \in \mathbf{x}$. Use Exercise 7.9 to show that $\hat{x}$ has only finitely many entries equal to zero.

7.12 Let $\hat{a}$ and $\hat{b}$ be equivalent Cauchy sequences, $\hat{a} \equiv \hat{b}$. Show that if $\hat{a}$ is positive, so is $\hat{b}$.

7.13 Let $\mathbf{x} = [\![\hat{a}]\!]$ and $\mathbf{y} = [\![\hat{b}]\!]$. Suppose that $a_n \leq b_n$ for all n. Show that $\mathbf{x} \leq \mathbf{y}$.

7.14 Let **a** be a nonzero real number. Show that one of **a** or $-\mathbf{a}$ is positive.

7.15 Let **a** and **b** be positive real numbers. Show that both $\mathbf{a} + \mathbf{b}$ and $\mathbf{a} \cdot \mathbf{b}$ are positive.

7.16 Let $\hat{a}$ be the sequence of left end points generated by the bisection algorithm in Figure 7.5. The values are given on page 116.

a) Write the values of the first several members of $\hat{a}$ exactly in binary, adding one extra digit for each member.

b) Write the value of $\sqrt{2}$ approximately in binary.

c) Describe the relationship of the successive terms of $\hat{a}$ in terms of the binary representation of $\sqrt{2}$.

7.17 The bisection method is a numerical technique for solving equations such as $x^2 - 2 = 0$. Each iteration of the bisection method gives better and better approximate solutions.

Newton's method (also known as the Newton–Raphson method) is another numerical technique for solving equations. For the equation $x^2 - 2 = 0$ the process is as follows:

- Start with an initial guess for the solution to the equation. Since we know the solution is somewhere between 1 and 2, let's say we start at $x = 1.5$.

- For each guess x, we calculate new guesses by the formula

$$x \leftarrow x - \frac{x^2 - 2}{2x}.$$

Compare the results of performing five iterations of the bisection method (starting with the interval $[1, 1.5]$) and five iterations of Newton's method (starting with initial guess 1.5).

Which yields a more accurate approximation for $\sqrt{2}$?

Chapter 8

ℝ: Real Numbers III, Complete Ordered Fields

We have two wholly different definitions of the real number system. In Chapter 6, real numbers were defined as Dedekind cuts. In Chapter 7, real numbers were defined as equivalence classes of Cauchy sequences. Neither of these is exactly the same as using the familiar decimal notation for real numbers.

Which of these is really the real numbers?

The answer is: They all are because, in a sense, they are all the same. That's an audacious claim because clearly Dedekind cuts look nothing like equivalence classes of Cauchy sequences. However, the thing both of these approaches have in common is they both define *complete ordered fields*. They are not the same, but they are *isomorphic*. Let's explore what that means.

8.1 Isomorphism

In mathematics, *isomorphic* objects are things that behave exactly the same as each other, but simply have different names. Let's look at some examples and then give a more precise definition in the context of the real number system.

The Fifteen Game

Let's play a game. The name of the game is *Fifteen* because the objective is to collect three number tiles that add up to exactly 15.

The game pieces are nine tiles numbered 1 through 9. Two players alternately choose a tile. The first player to hold three tiles that add up to 15 wins.

For example, suppose Player #1 starts by choosing tile 5. Player #2 responds by taking tile 3. Next Player #1 chooses tile 2. The state of the game at this point is illustrated in Figure 8.1.

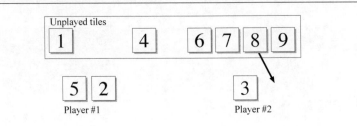

Figure 8.1: After three moves of the Fifteen game, Player #1 holds tiles 5 and 2, and Player #2 holds tile 3. To avoid a loss, Player #2 must choose tile 8.

In order not to lose, Player #2 must choose tile 8, because otherwise Player #1 would achieve $5 + 2 + 8 = 15$.

Now Player #1 must choose tile 4 for otherwise Player #2 would hold $8+3+4 = 15$. So after these moves the situation is this:

- Player #1 holds 2, 4, and 5,
- Player #2 holds 3 and 8.

What should Player #2 do? Since Player #1 holds 2 and 4, Player #2 must choose tile 9 to prevent #1 from winning with $2 + 4 + 9 = 15$. However, Player #1 also holds 4 and 5, so Player #2 needs to pick 6 to prevent $4 + 5 + 6 = 15$. Player #2 is doomed in either case, and Player #1 wins on the next move.

Have you played this game before? No?? I find that hard to believe because another name for this game is *tic-tac-toe*!

How is this tic-tac-toe? There's no three-by-three grid. Players don't use Xs and Os. There are no horizontal, vertical, or diagonal three-in-a-row configurations.

The connection between the Fifteen game and tic-tac-toe is demonstrated by a three-by-three arrangement of the numbers 1 through 9 in which the sum of the three numbers in any row, in any column, or in either diagonal is 15. Such an array of numbers is known as a *magic square*. See Figure 8.2.

Notice that the only three-element subsets of $\{1, 2, 3, \ldots, 9\}$ that add up to 15 are precisely these:

$$\{1,5,9\}, \quad \{1,6,8\}, \quad \{2,4,9\}, \quad \{2,5,8\},$$
$$\{2,6,7\}, \quad \{3,4,8\}, \quad \{3,5,7\}, \quad \{4,5,6\}.$$

These are exactly the triples from the rows, columns, and diagonals of the magic square. No other triples add up to 15. So when Player #1 chooses a tile in the

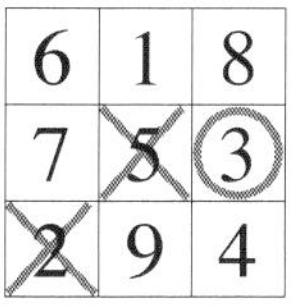

8   ℝ: Real Numbers III, Complete Ordered Fields

6	1	8
7	~~5~~	③
~~2~~	9	4

Figure 8.2: This three-by-three arrangement of the numbers 1 through 9 is a magic square because the row, column, and diagonal sums are all the same: 15. Choosing Xs and Os on this board exactly corresponds to choosing tiles in the Fifteen game. The game position shown here matches the position in Figure 8.1. The O player must choose the top right cell to block X from getting three in a row.

Fifteen game, that is exactly the same as placing an X on the tic-tac-toe board. Likewise, Player #2's tile choice is the same as marking an O.

The Fifteen game and tic-tac-toe are *isomorphic*. The possible moves and winning combinations in one game are in a precise one-to-one correspondence with the moves and winning combinations in the other. The difference between the two games is entirely superficial; at their core they are exactly the same game.

The Number Systems $(\mathbb{Z}_6, +)$ and $(\mathbb{Z}_7^*, \cdot)$

Next we consider an arithmetic example. Chapter 4 presents modular arithmetic, that is, the number system $(\mathbb{Z}_m, +, \cdot)$. The symbol $\mathbb{Z}_m$ stands for the set $\{0, 1, 2, \ldots, m-1\}$ and the operations are addition and multiplication modulo m.

We use the symbol $\mathbb{Z}^*$ to stand for the nonzero integers. In that spirit, we write $\mathbb{Z}_m^*$ to stand for the nonzero elements of $\mathbb{Z}_m$.

Here are the addition table for $\mathbb{Z}_6$ (left) and the multiplication table for $\mathbb{Z}_7^*$ (right).

+	0	1	2	3	4	5
0	0	1	2	3	4	5
1	1	2	3	4	5	0
2	2	3	4	5	0	1
3	3	4	5	0	1	2
4	4	5	0	1	2	3
5	5	0	1	2	3	4

·	1	2	3	4	5	6
1	1	2	3	4	5	6
2	2	4	6	1	3	5
3	3	6	2	5	1	4
4	4	1	5	2	6	3
5	5	3	1	6	4	2
6	6	5	4	3	2	1

At first glance these appear rather different. In addition to the different names for the elements (0 only on the left and 6 only on the right), the pattern of entries in the tables does not match. The addition table features "stripes" of numbers running from lower left to upper right (trace all the 4s in the left table); that pattern does not appear in the multiplication table.

However, if we list the elements of $\mathbb{Z}_7^*$ in the order $1, 3, 2, 6, 4, 5$ the tables look like this:

+	0	1	2	3	4	5
0	0	1	2	3	4	5
1	1	2	3	4	5	0
2	2	3	4	5	0	1
3	3	4	5	0	1	2
4	4	5	0	1	2	3
5	5	0	1	2	3	4

·	1	3	2	6	4	5
1	1	3	2	6	4	5
3	3	2	6	4	5	1
2	2	6	4	5	1	3
6	6	4	5	1	3	2
4	4	5	1	3	2	6
5	5	1	3	2	6	4

Notice that the striped pattern for both tables is the same. Let's write this down as a particular one-to-one correspondence between $\mathbb{Z}_6$ and $\mathbb{Z}_7^*$:

$\mathbb{Z}_6$		$\mathbb{Z}_7^*$
0	↔	1
1	↔	3
2	↔	2
3	↔	6
4	↔	4
5	↔	5

Notice that if we add numbers in $\mathbb{Z}_6$ the result corresponds to multiplying their mates in $\mathbb{Z}_7^*$. For example,

$$\begin{array}{ccccc} 3 & + & 4 & = & 1 \quad \text{in } \mathbb{Z}_6 \\ \updownarrow & & \updownarrow & & \updownarrow \\ 6 & \cdot & 4 & = & 3 \quad \text{in } \mathbb{Z}_7^* \end{array}$$

In other words, the structures of $(\mathbb{Z}_6, +)$ and $(\mathbb{Z}_7^*, \cdot)$ are exactly the same once we properly line up the elements of the two sets.

We checked only one example; how do we know this always works? Let's think of the elements of $\mathbb{Z}_7^*$ as powers of 3, like this:

$$3^0 = 1, \quad 3^1 = 3, \quad 3^2 = 2, \quad 3^3 = 6, \quad 3^4 = 4, \quad 3^5 = 5, \quad \text{and} \quad 3^6 = 1 = 3^0.$$

The earlier calculation $6 \cdot 4 = 3$ in $\mathbb{Z}_7^*$ can be rewritten as $3^3 \cdot 3^4 = 3^{3+4} = 3^7 = 3^1$. Here's the full correspondence:

$$\begin{array}{ccccc} 6 & \cdot & 4 & = & 3 \quad \text{in } \mathbb{Z}_7^* \\ \updownarrow & & \updownarrow & & \updownarrow \\ 3^3 & \cdot & 3^4 & = & 3^1 \quad \text{in } \mathbb{Z}_7^* \\ \updownarrow & & \updownarrow & & \updownarrow \\ 3 & + & 4 & = & 1 \quad \text{in } \mathbb{Z}_6 \end{array}$$

8.2 All Complete Ordered Fields are Isomorphic

Chapter 6 defines real numbers as Dedekind cuts and Chapter 7 defines real numbers as equivalence classes of Cauchy sequences. In both cases, the numbers (together with addition, multiplication, and less-than-or-equal) form a complete ordered field. Furthermore, there are other complete ordered fields besides these two (see [13]).

The good news is we don't have to decide which we prefer, thanks to the following important result.

Theorem 8.1. *Any two complete ordered fields are isomorphic.*

This means there is a one-to-one correspondence between the two fields that preserves the operations and the order relation. Let's explain that a bit further.

Suppose that elements a and b in one complete ordered field correspond to elements x and y in the other; that is $a \to x$ and $b \to y$. The correspondence "preserves" the operations if the following are true:

- $(a + b) \to (x + y)$,
- $(a \cdot b) \to (x \cdot y)$,
- $a \leq b$ exactly when $x \leq y$.

In other words, the two fields are exactly the same; only the way the elements of the fields are named is different.

The proof of Theorem 8.1 can be found in other books (e.g., [11]), but here is the gist.

We start by showing that the additive identities for the two fields must correspond to each other. Likewise for the multiplicative identities. In other words, both fields have a 0 and a 1 that correspond to each other.

Since the addition operation is preserved, $1 + 1$ in one fields exactly corresponds to $1 + 1$ in the other. So both fields have an element we can call 2. Continuing this way, we see that both fields contain a copy of the natural numbers, $\mathbb{N}$.

Next we show that additive inverses are preserved, so both fields contain the negatives of all the natural numbers. In this manner, both fields contain a copy of the full set of integers, $\mathbb{Z}$.

From there we show that multiplicative inverses are preserved, and this results in both fields containing copies of the rational numbers, $\mathbb{Q}$.

The final step is the most complicated and involves the use of the completeness property to show that every element in the first complete ordered field has a corresponding element in the other.

Theorem 8.1 is great news! It means we can forget whether we are using Dedekind cuts, or equivalence classes of Cauchy sequences, or decimal numbers, or some other exotic way of defining real numbers. At the end of the day, the real number system is a complete ordered field and it makes no difference how it is presented to us.

8.3 The Real Numbers have a Square Root of 2

Chapter 6 defines real numbers in terms of left rays. In Section 6.5 we construct a real number using this left ray

$$\{x \in \mathbb{Q}: x < 0\} \cup \{x \in \mathbb{Q}: x^2 < 2\}$$

and show that it is the square root of 2 (see page 92).

Chapter 7 defines real numbers in terms of Cauchy sequences. In Section 7.6 we use the bisection method to find a Cauchy sequence that represents the square root of 2.

Since all complete ordered fields are isomorphic, it must be the case that there is a real number whose square is 2 regardless of how we define the real number system. To that end, we show that the real numbers contain a $\sqrt{2}$ just using the fact that we are operating in a complete ordered field.

Proposition 8.2. *There is a real number r such that $r^2 = 2$.*

We begin by defining the following set:

$$A = \{a \in \mathbb{R}: a^2 \le 2\}.$$

The set A is nonempty; for example, $1 \in A$ because $1^2 = 1 < 2$.

The set A has an upper bound. For example, because $2^2 = 4 > 2$ it follows that every element in A is less than 2.

Therefore, by completeness, the set A has a least upper bound; let's call that number r. Note that $r \ge 1$.

We claim that $r^2 = 2$. To justify that claim, we rule out the possibility that $r^2 > 2$ and $r^2 < 2$.

- Suppose $r^2 > 2$. Let $t = r^2 - 2$; this is a positive number. Rewritten, this says $r^2 = 2 + t$.

  Let x be a real number with $0 < x < 1$; we'll be more specific in a moment. Because $r \ge 1$ and $x < 1$ we have that $2r - x > 2 - 1 = 1$. Therefore

  $$(r-x)^2 = r^2 - 2rx + x^2$$
  $$= (2+t) - x(2r-x)$$
  $$> (2+t) - x.$$

Now we specify x; let $x = t/2$. Then we continue the calculation to give

$$(r-x)^2 > (2+t) - x$$
$$= 2 + t - \frac{t}{2}$$
$$= 2 + \frac{t}{2} > 2.$$

It now follows that $r - x$ is also an upper bound for A, but it is smaller than r, contradicting the fact that r is the *least* upper bound for A. We have ruled out $r^2 > 2$.

- Suppose $r^2 < 2$. Then certainly we have $r < 2$. Let $t = 2 - r^2$; this is a positive number and we have $r^2 = 2 - t$. For $0 < x < 1$ we have

$$(r+x)^2 = r^2 + 2rx + x^2$$
$$= (2-t) + 2rx + x^2$$
$$= (2-t) + x(2r + x)$$
$$< (2-t) + 5x \qquad \text{because } r < 2 \text{ and } x < 1.$$

Letting $x = t/10$ we have

$$(r+x)^2 < (2-t) + 5x$$
$$= (2-t) + \frac{5t}{10}$$
$$= 2 - t + \frac{t}{2}$$
$$= 2 - \frac{t}{2} < 2.$$

This implies that $r + x \in A$ but $r + x$ is larger than r which is an upper bound for all of A. This is a contradiction and therefore we can rule out $r^2 < 2$.

Since neither $r^2 > 2$ nor $r^2 < 2$ it must be the case that $r^2 = 2$ and we have shown that there is a real number $\sqrt{2}$.

We have arrived at our final destination! The real numbers are a *complete ordered field*. It makes no difference which complete ordered field we call the real numbers because all complete ordered fields are identical (except for some cosmetic differences).

Kronecker simply took the integers as "given" and set forth working on mathematics from there. At this point, we can just take the real numbers as "given." We only need to know that there is a complete ordered field and, except for notation changes, there is only one complete ordered field. Everything we need to assume about the real numbers is embedded in those three words: *complete ordered field*.

Furthermore, the set of real numbers $\mathbb{R}$ contains a copy of the rational numbers, which, in turn, contains a copy of the integers, which contains a copy of the natural numbers:

$$\mathbb{N} \subseteq \mathbb{Z} \subseteq \mathbb{Q} \subseteq \mathbb{R}.$$

Each step of this journey repaired defects. The defects can all be expressed as an inability to solve certain equations for an unknown x:

- In the context of natural numbers, we can't always solve equations of the form $a + x = b$.

- In the context of integers, we can't always solve equations of the form $ax = b$ (where $a \neq 0$).

- In the the context of rational numbers, we can't always solve equations of the form $x^2 = a$.

Can we always solve equations of the form $x^2 = a$ in the context of the real numbers? Alas, as you know, the answer is yes only if $a \geq 0$; we can't solve this equation: $x^2 = -1$. We have more work to do.

Recap

Chapters 6 and 7 presented two different definitions of real numbers: equivalence classes of left rays and equivalence classes of Cauchy sequences of rational numbers. Which is the "correct" definition of real numbers?

This chapter is a reconciliation. In both instances, we found that the real numbers are complete ordered fields. We introduced the concept of isomorphism and then asserted that any two complete ordered fields must be isomorphic. Hence the two seemingly disparate constructions are simply different versions of the same structure.

The real numbers are a complete ordered field. Which complete ordered field? It doesn't matter; they are all the same. That says it all.

Exercises

8.1 The rational numbers $\mathbb{Q}$ and the real numbers $\mathbb{R}$ are ordered fields. One might wonder: Are there other ordered fields? Here is one that is nestled between $\mathbb{Q}$ and $\mathbb{R}$.

Let

$$\mathbb{Q}\left[\sqrt{2}\right] = \left\{a + b\sqrt{2} : a, b \in \mathbb{Q}\right\}.$$

For example, $\frac{1}{2} - \frac{3}{2}\sqrt{2}$ is an element of $\mathbb{Q}\left[\sqrt{2}\right]$.

a) Show that the sum and product of elements of $\mathbb{Q}\left[\sqrt{2}\right]$ are also elements of $\mathbb{Q}\left[\sqrt{2}\right]$.

b) Let $a, b \in \mathbb{Q}$ and suppose they are not both zero. Show that $a + b\sqrt{2}$ has a multiplicative inverse that is an element of $\mathbb{Q}\left[\sqrt{2}\right]$.

c) Give a criterion on rational numbers a and b for when $a + b\sqrt{2} > 0$.

8.2 Find an isomorphism between $(\mathbb{Z}_{11}^*, \cdot)$ and $(\mathbb{Z}_{10}, +)$ by considering powers of 2 in $\mathbb{Z}_{11}^*$.

Explain why considering powers of 3 in $\mathbb{Z}_{11}^*$ does not yield an isomorphism.

8.3 Crudely speaking, any mathematical thing is isomorphic to itself. We call an isomorphism of a thing with itself an *automorphism*.

For example, consider $(\mathbb{Z}_{10}, +)$. An easy isomorphism is to associate each element of $\mathbb{Z}_{10}$ with itself: $0 \leftrightarrow 0$, $1 \leftrightarrow 1$, $2 \leftrightarrow 2$, and so forth.

Formally, we define the function f from $\mathbb{Z}_{10}$ to $\mathbb{Z}_{10}$ simply as $f(a) = a$ for all $a \in \mathbb{Z}_{10}$. This is a one-to-one correspondence with the property that $f(a + b) = f(a) + f(b)$. Alas, this is not interesting at all.

Here is a different association that's more interesting. Pair $1 \leftrightarrow 3$. That is, define a new function f in which $f(1) = 3$. Now figure out how to extend this function so that $f(a + b) = f(a) + f(b)$ for $\mathbb{Z}_{10}$. This gives an interesting automorphism of $(\mathbb{Z}_{10}, +)$.

Finally, show that defining yet another function g with $g(1) = 2$ cannot be extended to an automorphism of $(\mathbb{Z}_{10}, +)$.

8.4 Show that the sequence

$$\hat{a} = \left(\frac{1}{\sqrt{1}}, \frac{1}{\sqrt{2}}, \frac{1}{\sqrt{3}}, \frac{1}{\sqrt{4}}, \ldots\right)$$

is a Cauchy sequence of real numbers. (In Chapter 7 we considered Cauchy sequences of rational numbers only, but the concept extends to real numbers.)

The following exercises pertain to standard decimal (base-ten place-value) notation for real numbers.

8.5 A natural number can be expressed as a finite sequence $s_1 s_2 s_3 \ldots s_n$ where each s_i is one of the ten symbols 0, 1, 2, 3, 4, 5, 6, 7, 8, and 9. These ten symbols are called *digits*.

Give rules that describe which sequences are properly formed natural numbers so that there is only one representation for each natural number.

8.6 Like natural numbers, integers may be also be expressed as a finite sequence $s_1 s_2 \ldots s_n$, but in this case the symbols may be one of eleven possibilities: a minus sign or a digit.

Give rules that describe which sequences are properly formed integers so that there is only one representation for every integer.

8.7 Real numbers can be expressed in decimal notation, but now the sequences may be infinitely long, comprised of the twelve symbols 0 through 9, the minus sign (−) and the decimal marker (either . or , depending on region).

Which sequences of these twelve symbols are proper representations of real numbers?

8.8 In the previous exercise, you were asked to describe which sequences of symbols are proper decimal representations of real numbers. Standard notation allows different sequences to represent the same real number, giving an equivalence relation on decimal notations. For example, all of the following are equivalent representations of the real number 1:

$$1 \quad 1. \quad 1.0 \quad 1.00 \quad 1.000\ldots \quad 0.9999\ldots$$

The number one-tenth has all of these possible decimal representations:

$$.1 \quad .10 \quad .1000\ldots \quad 0.1 \quad 0.10 \quad 0.1000\ldots \quad .9999\ldots \quad 0.9999\ldots$$

Give rules that explain when two decimal notations are equivalent, that is, when the sequences represent the same real number.

8.9 Another way to represent real number is using *scientific notation*. For example, the real number 123.456 is properly written as 1.23456×10^2. However, none of the following are valid[1] ways to write this number: 12.3456×10^1 or 1234560×10^{-4} or 0.123456×10^3.

Give rules that specify when a number is properly written in scientific notation. How is 0 expressed in your system?

8.10 Let $x = 5.3520153\ldots$ and $y = 2.498151\ldots$ where we have shown only the first few digits of these real numbers. Is it possible to determine the first two digits to the right of the decimal point for $x + y$ and for $x \cdot y$? If so, what are they?

Addendum: Cauchy Sequences and Completeness Revisited

In Chapter 7 we define real numbers as equivalence classes of Cauchy sequences of *rational* numbers. Definition 7.1 requires the elements of the sequence to be

---

[1] Or, at least, not standard.

rational, but once real numbers have been defined, the definition makes sense for a sequence of *real* numbers.

For example, consider this sequence:

$$\hat{a} = \left(\tfrac{1}{\sqrt{1}}, \tfrac{1}{\sqrt{2}}, \tfrac{1}{\sqrt{3}}, \tfrac{1}{\sqrt{4}}, \ldots\right).$$

Only some of the elements of this sequence are rational. Nevertheless, it is a Cauchy sequence; see Exercise 8.4.

Furthermore, we can apply Definition 7.2 and note that $\hat{a}$ is equivalent to the sequence $(0, 0, 0, \ldots)$ because the difference between their nth terms is $1/\sqrt{n}$, which is always less than any given $s > 0$ once $n > s^2$.

We say that the sequence $\hat{a}$ *converges* to the real number 0. Here is the general definition.

Definition 8.3. Let $\hat{a}$ be a Cauchy sequence and let ℓ be a number. We say that $\hat{a}$ *converges* to ℓ provided $\hat{a}$ is equivalent to the sequence $(\ell, \ell, \ell, \ldots)$.

In the notation of calculus, we may write this as

$$\lim_{n \to \infty} a_n = \ell.$$

We define *completeness* of the real numbers using the concept of least upper bound. While it is beyond the scope of this book, it is worth noting that one can prove that all Cauchy sequences of real numbers converge to some value using the fact that all nonempty sets of real numbers that have an upper bound have a least upper bound.

Furthermore, any ordered field in which Cauchy sequences converge must satisfy the least upper bound property. The concept of *completeness* is usually defined in terms of the convergence of Cauchy sequences.

In other words, an ordered field satisfies the least upper bound property if and only if all of its Cauchy sequences converge. This allows us to declare other fields complete even if they are not ordered. In the next chapter we introduce the complex numbers. They too form a field, but they are not ordered. Still, the complex numbers are a complete field because Cauchy sequences of complex numbers also converge.

Chapter 9

ℂ: Complex Numbers

Just as the natural numbers, integers, and rational numbers are all embedded inside the real numbers, the real numbers in turn are embedded in a larger set of numbers: the *complex* numbers ℂ.

Each step of the journey from natural to real numbers was motivated by our inability to solve certain equations. The rational numbers are rich enough to solve simple linear equations of the form $ax + b = c$ (provided $a \neq 0$) but are not sufficient to solve quadratic equations such as $x^2 - 2 = 0$.

The real numbers are only a partial solution to this conundrum; we can solve equations of the form $x^2 - a = 0$ only when $a \geq 0$. However, there is no real number x such that $x^2 + 1 = 0$.

9.1 From i to ℂ

The rational numbers ℚ and the real numbers ℝ are fields. This means that all the familiar properties of addition and multiplication are valid.

The central idea in solving the equation $x^2 + 1 = 0$ is simply to invent a new number called i with the property that $i^2 = -1$. That solves that particular equation. However, it is desirable that all the usual rules of algebra are still available to us. In other words, we still want to be able to add, subtract, multiply, and divide with this new number.

To do that, we need to be able to multiply i by real numbers and add real numbers to the result. To extend the real numbers ℝ by the incorporation of i as well as to maintain all the familiar rules of algebra leads to the following definition of *complex numbers*.

Definition 9.1. A *complex number* is a number of the form $a + bi$ where a and b are real numbers. The set of all complex numbers is denoted ℂ.

The definitions of addition and multiplication nearly write themselves.

> **Definition 9.2.** Addition and multiplication of complex numbers are defined as follows:
> $$(a + bi) + (c + di) = (a + c) + (b + d)i,$$
> $$\begin{aligned}(a + bi) \cdot (c + di) &= a(c + di) + bi(c + di)\\ &= ac + adi + bci + bdi^2 \\ &= (ac - bd) + (ad + bc)i.\end{aligned}$$

From here it is easy to check that $0+0i$ is the identity element for addition and $1 + 0i$ is the identity element for multiplication. The complex number $a + bi$ has an additive inverse, namely $(-a) + (-b)i$, which we may also write as $-a - bi$.

Finally, let $a + bi$ be a complex number different from $0 + 0i$; we need to show that it has a multiplicative inverse. We claim that the following number is that inverse:

$$\left(\frac{a}{a^2 + b^2}\right) + \left(\frac{-b}{a^2 + b^2}\right)i.$$

Notice that the denominator, $a^2 + b^2$, is not zero because the assumption that $a + bi \neq 0 + 0i$ implies that one (or both) of a or b is not zero.

To check that this is, indeed, the multiplicative inverse of $a + bi$ we multiply and see what happens:

$$\begin{aligned}[a + bi] \cdot &\left[\left(\frac{a}{a^2 + b^2}\right) + \left(\frac{-b}{a^2 + b^2}\right)i\right]\\ &= \left(\frac{a^2}{a^2 + b^2} - \frac{-b^2}{a^2 + b^2}\right) + \left(\frac{-ab}{a^2 + b^2} + \frac{ba}{a^2 + b^2}\right)i \\ &= \left(\frac{a^2 + b^2}{a^2 + b^2}\right) + \left(\frac{-ab + ba}{a^2 + b^2}\right)i \\ &= 1 + 0i.\end{aligned}$$

It is routine to check that all of the usual algebraic properties enjoyed by $\mathbb{Q}$ and $\mathbb{R}$ are also upheld in $\mathbb{C}$ and we have the following.

> **Proposition 9.3.** *The complex number system* $(\mathbb{C}, +, \cdot)$ *is a field.*

Notice that a copy of the real numbers $\mathbb{R}$ is embedded in $\mathbb{C}$; the reals are simply the numbers of the form $a + 0i$. We therefore think of $\mathbb{R}$ as a proper subset of $\mathbb{C}$ and we have this grand string of inclusions:

$$\mathbb{N} \subseteq \mathbb{Z} \subseteq \mathbb{Q} \subseteq \mathbb{R} \subseteq \mathbb{C}.$$

No Order

All of the previous number systems had, in addition to the operations of addition and multiplication, the ordering relation $\leq$. Can we extend the definition of $\leq$ to the complex numbers?

The answer is mostly no. We can define a relation $\leq$ for the complex numbers (see Exercise 9.4) but there's no good way to give an ordering that plays nicely with addition and multiplication. To see why, suppose we had complex numbers w and z that are both greater than zero. Then we would want both $w + z$ and $w \cdot z$ to be greater than zero (as in Proposition 5.9). In particular, this implies that if $z \neq 0$, then z^2 must be positive; but we know that $i^2 < 0$.

9.2 Complex Numbers as Equivalence Classes

A recurring motif in this book is to define numbers as equivalence classes. Let's see how we can apply this paradigm to create complex numbers.

Recall from Chapter 4 that we created modular numbers $\mathbb{Z}_m$ using the full set of integers as the raw ingredients. Then two integers are equivalent if their difference is divisible by m and $\mathbb{Z}_m$ consists of the equivalence classes.

In this case the raw ingredients are polynomials with real coefficients. Let $\mathbb{R}[x]$ denote the set of all polynomials of the form

$$a_n x^n + a_{n-1} x^{n-1} + \cdots + a_1 x + a_0$$

where the coefficients are real numbers. (This is analogous to $\mathbb{Z}[x]$ introduced in Exercise 3.18.)

Next we define a relation on $\mathbb{R}[x]$. We say that polynomials $p(x)$ and $q(x)$ are equivalent provided $p(x) - q(x)$ is divisible by $x^2 + 1$. It is most proper to write this as

$$p(x) \equiv q(z) \pmod{x^2 + 1},$$

but since this is the only equivalence relation we are considering in this chapter, we may omit the $(\bmod\, x^2 + 1)$ part of the notation.

Proposition 9.4. *The relation* $(\bmod\, x^2 + 1)$ *for polynomials is an equivalence relation.*

You are asked to prove this in Exercise 9.5.

It is useful to consider some examples. Let

$$p(x) = 2x^3 - 3x^2 + 5x \quad \text{and} \quad q(x) = x^3 - 2x^2 + 4x + 1.$$

Subtracting and factoring we have

$$p(x) - q(x) = x^3 - x^2 + x - 1 = (x-1)(x^2+1).$$

Since $p(x) - q(x)$ is a multiple of $x^2 + 1$ we have $p(x) \equiv q(x)$.

Here's another example. Let's start with the polynomial x^3. If we add or subtract a multiple of $x^2 + 1$ the result will be equivalent. Let's subtract $x(x^2+1) = x^3 + x$:

$$x^3 - [x(x^2+1)] = x^3 - (x^3 + x) = -x$$

and we conclude that $x^3 \equiv -x$.

Notice that the new, equivalent polynomial $(-x)$ has a lower degree than the original (x^3). Let's try a more elaborate example.

Let $p(x) = x^4 - 3x^2 + 5x + 2$. When we add or subtract a multiple of $x^2 + 1$, the result will be equivalent. We start by eliminating the x^4 term. To do that, we subtract $x^2(x^2 + 1) = x^4 + x^2$. Let

$$q(x) = p(x) - x^2(x^2+1) = \left(x^4 - 3x^2 + 5x + 2\right) - \left(x^4 + x^2\right)$$
$$= -4x^2 + 5x + 2.$$

Because $p(x)$ and $q(x)$ differ by a multiple of $x^2 + 1$, we have $p(x) \equiv q(x)$.
Now we remove the $-4x^2$ term in $q(x)$ by adding $4(x^2 + 1)$. Let

$$r(x) = p(x) + 4(x^2+1) = \left(-4x^2 + 5x + 2\right) + \left(4x^2 + 4\right) = 5x + 6.$$

What we see is that, given any polynomial of degree 2 or higher, we can add (or subtract) multiples of $x^2 + 1$ to find an equivalent polynomial of the form $ax + b$. Further, the only way for $ax + b$ to be equivalent to $cx + d$ is if $a = c$ and $b = d$. We summarize this here:

Proposition 9.5. *Let $p(x) \in \mathbb{R}[x]$. There is a unique polynomial of the form $ax + b$ such that $p(x) \equiv ax + b$.*

We now define the complex numbers:

9   C: Complex Numbers

Definition 9.6. The complex numbers, $\mathbb{C}$, are the equivalence classes of $\mathbb{R}[x]$ for the relation $(\bmod\ x^2 + 1)$.

Addition and multiplication are defined by

$$[\![p(x)]\!] + [\![q(x)]\!] = [\![p(x) + q(x)]\!],$$
$$[\![p(x)]\!] \cdot [\![q(x)]\!] = [\![p(x) \cdot q(x)]\!].$$

Where is i in this definition? Because $x^2 \equiv -1$, we have $[\![x]\!] \cdot [\![x]\!] = [\![x^2]\!] = [\![-1]\!]$. In other words, $i = [\![x]\!]$. More generally, the complex number $a + bi$ is precisely $[\![a + bx]\!]$. Let's check that this works correctly.

Let **w** and **z** be complex numbers: say, $\mathbf{w} = a+bi = [\![a+bx]\!]$ and $\mathbf{z} = c+di = [\![c+dx]\!]$. Let's consider their sum and product.

$$\begin{aligned}
\mathbf{w} + \mathbf{z} = (a+bi) + (c+di) &= [\![a+bx]\!] + [\![c+dx]\!] \\
&= [\![(a+bx) + (c+dx)]\!] \\
&= [\![(a+c) + (b+d)x]\!] \\
&= (a+c) + (b+d)i
\end{aligned}$$

$$\begin{aligned}
\mathbf{w} \cdot \mathbf{z} = (a+bi) \cdot (c+di) &= [\![a+bx]\!] \cdot [\![c+dx]\!] \\
&= [\![(a+bx) \cdot (c+dx)]\!] \\
&= [\![ac + bdx + adx + bcx^2]\!] \\
&= [\![ac - bd + adx + bcx]\!] \quad \text{because } x^2 \equiv -1 \\
&= [\![(ac - bd) + (ad + bc)x]\!] \\
&= (ac - bd) + (ad + bc)i.
\end{aligned}$$

This shows that whether we use Definition 9.2 or Definition 9.6 we arrive at exactly the same result.

9.3   Polar Coordinates

Real numbers can be visualized as a number line. The number a is positioned distance $|a|$ away from 0 on the line either to the right (if a is positive) or to the left (if a is negative).

Complex numbers also have a geometric interpretation. The complex number $a + bi$ can be visualized as a point in a plane at coordinates (a, b).

Addition of complex numbers behaves exactly like vector addition. The complex number $a + bi$ is located a units to the right and b units up from the origin. Adding $c + di$ takes us an additional c units to the right and d units vertically. See Figure 9.1.

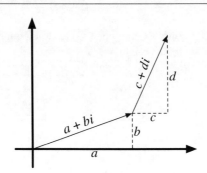

Figure 9.1: This is a visualization of the addition of the complex numbers $a + bi$ and $c + di$.

There is also an attractive way to visualize complex multiplication, but to understand it we need to begin with the introduction of *polar coordinates*. Instead of describing the complex number's location as distances horizontally and vertically, we specify the location by giving a distance from the origin, r, and a direction θ measured counterclockwise from the positive x-axis. We write (r, θ) as the polar coordinates of the complex number. See Figure 9.2.

When we examine the right triangle with leg lengths a and b and hypotenuse r, we have (using the Pythagorean Theorem and basic trigonometry) the four relations given in Proposition 9.7.

Proposition 9.7. *Let (r, θ) be the polar coordinates of the complex number $a + bi$. Then*
$$r = \sqrt{a^2 + b^2}, \qquad a = r\cos\theta,$$
$$\tan\theta = \frac{b}{a}, \qquad b = r\sin\theta.$$

The number r is called the *magnitude* and the angle θ is called the *phase* of the complex number with polar coordinates (r, θ).

Polar coordinates help us uncover the geometry of complex multiplication. Suppose we have two complex numbers with polar coordinates as follows:
$$z = a + bi \leftrightarrow (r, \theta) \quad \text{and} \quad w = c + di \leftrightarrow (s, \phi),$$

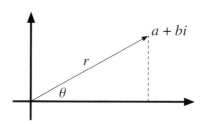

Figure 9.2: Representing the complex number $a + bi$ in polar coordinates (r, θ). Note that $r = \sqrt{a^2 + b^2}$ and $\tan \theta = b/a$. Conversely $a = r \cos \theta$ and $b = r \sin \theta$.

which gives
$$a = r \cos \theta, \quad c = s \cos \phi,$$
$$b = r \sin \theta, \quad d = s \sin \phi.$$

We derive the polar coordinates of $z \cdot w$ using some algebra and the angle-sum formulas for sine and cosine:

$$\begin{aligned} z \cdot w &= (a + bi)(c + di) \\ &= (ac - bd) + (ad + bc)i \\ &= \Big((r \cos \theta)(s \cos \phi) - (r \sin \theta)(s \sin \phi)\Big) \\ &\quad + \Big((r \cos \theta)(s \sin \phi) + (r \sin \theta)(s \cos \phi)\Big)i \\ &= (rs)\Big[(\cos \theta \cos \phi - \sin \theta \sin \phi) + (\cos \theta \sin \phi - \sin \theta \cos \phi)i\Big] \\ &= (rs)\Big[\cos(\theta + \phi) + \sin(\theta + \phi)i\Big]. \end{aligned}$$

The last line in polar coordinates is neatly expressed as $(rs, \theta + \phi)$.

Proposition 9.8. *Let (r, θ) and (s, ϕ) be the polar coordinates of two complex numbers. In polar coordinates, their product is given by*

$$(r, \theta) \cdot (s, \phi) = (rs, \theta + \phi).$$

In words, when we multiply two complex numbers z and w, the magnitude of the result is the product of magnitudes of z and w, and the phase of the result is the sum of the phases of z and w. See Figure 9.3.

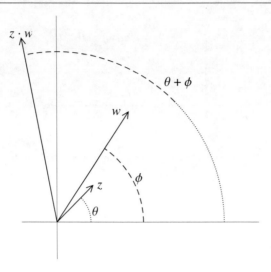

Figure 9.3: This is a visualization of complex multiplication of z with polar coordinates (r, θ) and w with polar coordinates (s, ϕ). The product has magnitude rs and phase $\theta + \phi$.

More Roots

The impetus to extend $\mathbb{R}$ to form $\mathbb{C}$ was the lack of a square root of -1. Not only do we find a $\sqrt{-1}$ in $\mathbb{C}$, but we can find square roots of any real number. For example, the square roots of -4 are $2i$ and $-2i$. More generally, if b is a positive real number then $\pm i\sqrt{b}$ are the square roots of $-b$.

But now we have more numbers. Do we need to extend $\mathbb{C}$ to find a square root of i? Fortunately not! Observe the following calculation:

$$\left(\frac{\sqrt{2}}{2} + \frac{\sqrt{2}}{2}i\right)^2 = \left(\frac{\sqrt{2}}{2} \cdot \frac{\sqrt{2}}{2} - \frac{\sqrt{2}}{2} \cdot \frac{\sqrt{2}}{2}\right) + \left(\frac{\sqrt{2}}{2} \cdot \frac{\sqrt{2}}{2} + \frac{\sqrt{2}}{2} \cdot \frac{\sqrt{2}}{2}\right)i$$

$$= 0 + \left(\frac{1}{2} + \frac{1}{2}\right)i$$

$$= i.$$

At first glance it might seem that finding this square root of i must be difficult but Proposition 9.8 helps enormously. The polar form of i is $(1, \frac{\pi}{2})$. To find a square root of i in polar form we want

$$z^2 = (r, \theta) \cdot (r, \theta) = (r^2, 2\theta) = (1, \tfrac{\pi}{2}).$$

From this we see that we need $r^2 = 1$ and $2\theta = \frac{\pi}{2}$ yielding $r = 1$ and $\theta = \frac{\pi}{4}$. Then

$$z = (r\cos\theta) + (r\sin\theta)i = (\cos\tfrac{\pi}{4}) + (\sin\tfrac{\pi}{4})i = \frac{\sqrt{2}}{2} + \frac{\sqrt{2}}{2}i.$$

There's nothing special about finding the square of i; we can use the same method for any complex number. If the polar coordinates of z are (r, θ), then $(\sqrt{r}, \theta/2)$ is a square root of z.

Of course, $(r, \theta + 2\pi)$ is also a polar form for z since adding a multiple of 2π does not change the angle. Working from this version, we find that $(\sqrt{r}, \pi + \theta/2)$ is also a square root of z; indeed, it's simply the negative of $(\sqrt{r}, \theta/2)$.

What about cube roots? The same method works fine. If $z = (r, \theta)$ then $(\sqrt[3]{r}, \theta/3)$ is a cube root of z, as are $(\sqrt[3]{r}, \theta/3 + 2\pi/3)$ and $(\sqrt[3]{r}, \theta/3 + 4\pi/3)$.

This idea can be extended to the nth root of any complex number (where n is a positive integer). In other words, for any complex number z, the equation $x^n - z = 0$ can be solved in $\mathbb{C}$.

The equation $x^n - z = 0$ is a rather special type of equation. There are, of course, many other kinds of polynomial equations such as

$$(3 - 2i)x^5 - 4x^2 + (1 - i)x + 5 = 0. \qquad (*)$$

Does such an equation have a solution in $\mathbb{C}$? The fabulous news is that the answer is yes and that is the subject of the next section.

9.4 The Fundamental Theorem of Algebra

A primary motivation for extending the real numbers to the complex numbers is to be able to solve equations of the form $x^2 + a = 0$. Since every complex number has a complex square root, we have accomplished this task. Furthermore, our ability to take square roots enables us to find a complex solution to the general quadratic equation $ax^2 + bx + c = 0$ where $a, b, c \in \mathbb{C}$ (and $a \neq 0$); we simply apply the quadratic formula:

$$\frac{-b \pm \sqrt{b^2 - 4ac}}{2a}.$$

What about cubic equations or other higher order polynomial equations? Let's see that all of these have solutions as well.

Drawing Curves in the Complex Plane

Consider the set of complex numbers with polar coordinates $(1, \theta)$; what does this set look like when plotted as points in the complex plane?

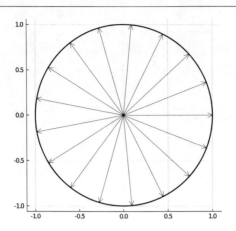

Figure 9.4: The set of all complex numbers with polar coordinates $(1, \theta)$, for $0 \le \theta \le 2\pi$, is a circle of radius 1 centered at $0 + 0i$. This set is named C_1; in general C_r is the set of all complex numbers with polar coordinates (r, θ).

Recall that $(1, \theta)$ is a complex number that's distance 1 from the origin, $0 + 0i$, at an angle of θ from the positive horizontal axis. As θ ranges from 0 to 2π, we sweep out all the points on a circle around the origin. See Figure 9.4.

More generally, for a nonnegative real number r let

$$C_r = \{(r, \theta) : 0 \le \theta \le 2\pi\}.$$

The set C_r is a circle of radius r centered at the origin, $0 + 0i$.

Let p be a polynomial. We define the set $p(C_r)$ to be the set of values we get when we apply p to every element of C_r. We look at some examples.

- Let $p(x) = x + 5$. What is $p(C_1)$? Recall that C_1 is the set of all complex numbers with polar coordinates $(1, \theta)$. In standard notation, this is $\cos \theta + i \sin \theta$. When we evaluate the polynomial at $x = (1, \theta) = \cos \theta + i \sin \theta$, the result is $(5 + \cos \theta) + (\sin \theta) i$. The set $p(C_1)$ is a circle of radius 1, but now it is centered at $5 + 0i$.

  The set $p(C_r)$ is a circle of radius r centered at $5 + 0i$.

- Let $p(x) = x^2$. What is $p(C_1)$? Recall that the square of a complex number with polar coordinates (r, θ) is $(r^2, 2\theta)$. In this case $r = 1$, so r^2 is also 1. Therefore $p(C_1)$ is the set of all complex numbers with polar coordinates $(1, 2\theta)$ for $0 \le \theta \le 2\pi$.

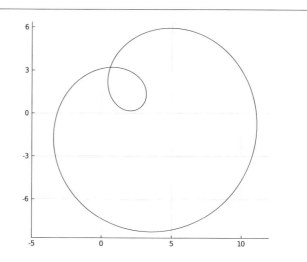

Figure 9.5: The curve $p(C_2)$ where p is the polynomial $x^2 + (2-i)x + 3$.

Said differently, $p(C_1)$ is the set $\{(1, \theta): 0 \le \theta \le 4\pi\}$. Of course, $(1, \theta) = (1, \theta + 2\pi)$ so this is a bit redundant, but the redundancy is interesting. If we imagine drawing the set $p(C_1)$ by tracing the points $p(z)$ for $z \in C_1$, we end up going around the unit circle twice. Of course, the end result looks exactly the same as C_1.

For other values of r, $p(C_r)$ is a circle of radius r^2 centered at $0 + 0i$.

- Let $p(x) = x^2 + (3-2i)$. Then $p(C_r)$ is a circle of radius r^2 centered at $3 - 2i$. The effect of the added term is simply to shift the center of the circle to a new location.

- Let $p(x) = x^2 + (2-i)x + 3$. The set $p(C_2)$ is the lovely curve shown in Figure 9.5.

A Close Look at a Particular Polynomial

We now focus on the polynomial from equation (∗) on page 143:

$$p(x) = (3-2i)x^5 - 4x^2 + (1-i)x + 5.$$

We plan to examine the curves $p(C_r)$ when r is a large number and when r is nearly zero.

Let x be a complex number (r, θ) with a large value of r. The x^5 term evaluates to $(r^5, 5\theta)$. If that were the only term in the polynomial, the curve

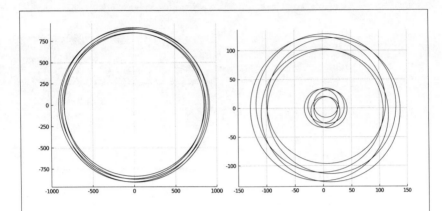

Figure 9.6: The curves $p(C_r)$ for $r = 3$ (left), $r = 2$ (right outer), and $r = 1.5$ (right inner) for the polynomial $p(x) = (3 - 2i)x^5 - 4x^2 + (1 - i)x + 5$.

$p(C_r)$ would be a circle that, as θ goes from 0 to 2π, would encircle the origin five times. The other terms matter, but because the exponent on r is smaller, the effect is to deflect the curve gently from its five-fold orbit around the origin. The left portion of Figure 9.6 shows the curve $p(C_r)$ when $r = 3$. (Note that $3^5 = 243$ is so much bigger than $3^2 = 9$ that when $r = 3$ the curve $p(C_r)$ is nearly a five-fold circle around the origin.) The right portion of the figure shows the curves $p(C_2)$ and, nestled inside, $p(C_{1.5})$. Notice, by looking at the axis markings, that the curves on the right are much smaller than $p(C_3)$. Also observe that the effects of the lower-order terms make the deviations from circular more pronounced.

As we decrease r further, the curves get smaller and smaller. Figure 9.7 shows the curve $p(C_{0.5})$. Notice that the curve winds around the point $p(0) = 5 + 0i$. Further, whereas the previous curves $p(C_r)$ we examined included $0 + 0i$ in their interior, for this curve we find the origin outside. Decreasing r further would simply make the curve wind more tightly around $p(0) = 5 + 0i$ because the other terms in the polynomial involve powers of r and those terms are getting smaller and smaller.

When $r = 1.5$ the curve $p(C_r)$ surrounds the origin, but when $r = 0.5$ the origin is on the outside. Therefore, for some value of r between 0.5 and 1.5 the curve exactly hits $0 + 0i$; this is illustrated in Figure 9.8.

What does it mean that the curve $p(C_r)$ passes through the origin, $0 + 0i$? It means that $p((r, \theta)) = 0 + 0i$ for some value of θ. In other words, (r, θ) is a root of this polynomial!

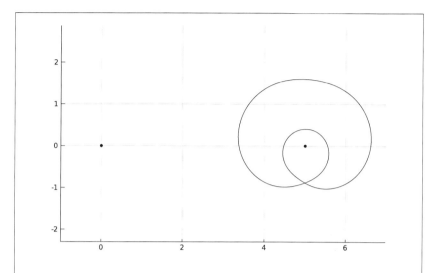

Figure 9.7: The curve $p(C_{0.5})$ for the polynomial $p(x) = (3-2i)x^5 - 4x^2 + (1-i)x + 5$. The origin, $0 + 0i$, and the point $p(0) = 5 + 0i$ are indicated with dots.

Closure

We have shown that there is a complex number z with the property that $p(z) = 0$ for the polynomial $p(x) = (3 - 2i)x^5 - 4x^2 + (1 - i)x + 5$. There is nothing special about this polynomial. We have the following important result:

Theorem 9.9 (Fundamental Theorem of Algebra). *Let p be a (non-constant) polynomial with complex coefficients. There is a complex number z such that $p(z) = 0$.*

Here's the argument. Let

$$p(x) = a_n x^n + a_{n-1} x^{n-1} + \cdots + a_1 x + a_0$$

where $n > 0$ and $a_n \neq 0$.

If $a_0 = 0 + 0i$, then naturally $p(0 + 0i) = 0 + 0i$ and so $z = 0 + 0i$ is a root of the equation $p(x) = 0$.

Otherwise, assume that $a_0 \neq 0 + 0i$. When r is large, the curve $p(C_r)$ is a wobbly circle that orbits the origin n times. When r is a small positive number, $p(C_r)$ is a curve tightly wrapped around a_0 and therefore the origin, $0 + 0i$, is

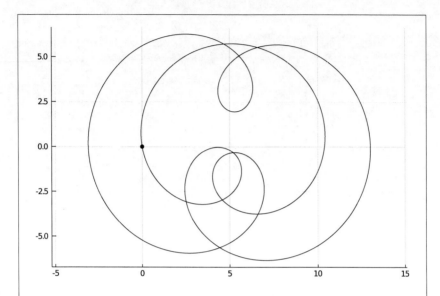

Figure 9.8: The curve $p(C_r)$ with $r = 0.98115381\ldots$ for the polynomial $p(x) = (3 - 2i)x^5 - 4x^2 + (1 - i)x + 5$. Observe that the $0 + 0i$ (shown with a dot) lies on this curve.

outside this curve. Hence, for some value of r between these extremes, the curve $p(C_r)$ goes through the origin. For this r, there is a θ so that $z = (r, \theta)$ satisfies $p(z) = 0 + 0i$.

As we progressed from natural numbers through to the reals we found simple equations that had no solutions. Finally, with the complex numbers, all polynomial equations have solutions. We express this by saying that the complex numbers are *algebraically closed*.

In this way, our journey, which began with the natural numbers, has a natural end with the complex numbers.

Recap

We introduced the number i with the property that $i^2 = -1$. Adjoining this to $\mathbb{R}$ and respecting the rules of algebra led us to the complex numbers $\mathbb{C}$. Alternatively, the complex numbers may be defined as equivalence classes of polynomials.

We find that not only can we solve all equations of the form $x^2 + a = 0$ for any complex number a, we can solve all polynomial equations $p(x) = 0$ in $\mathbb{C}$.

Exercises

9.1 The absolute value of a complex number $z = a + bi$ is defined to be $|z| = \sqrt{a^2 + b^2}$. Notice that $|z|$ is the distance from the origin, $0 + 0i$, to z in the complex plane.

Show that for complex numbers w and z we have $|w \cdot z| = |w| \cdot |z|$.

9.2 Suppose the complex number z is given in polar coordinates as (r, θ). Find $|z|$.

9.3 We know that $\sqrt[3]{1} = 1$, however there are two other complex numbers z with $z^3 = 1$. Find them.

9.4 The complex numbers $\mathbb{C}$ do not have a natural ordering, but it is possible to define one as follows. Suppose $w = a + bi$ and $z = c + di$ are complex numbers. We compare them by first seeing which has a larger real part; that is, is $a < c$ (in which case we say that w is smaller) or is $a > c$ (in which case we say that z is smaller). If $a = c$, then we compare using the imaginary parts. If $b < d$ we say w is smaller but if $b > d$ then we say z is smaller.

Since this is a nonstandard notion of less-than, we use the symbol $\prec$ to indicate this relation, which we call *lexicographic ordering*. Formally, the definition is that $(a + bi) \prec (c + di)$ provided either (1) $a < c$ or (2) $a = c$ and $b < d$.

The relation $\preceq$ imbues the complex numbers with an ordering, and yet we do not say that $\mathbb{C}$ is an ordered field. Why not?

Specifically, find complex numbers w and z with $w \succ 0$ and $z \succ 0$ but $w \cdot z \not\succ 0$.

9.5 Show that the relation $(\bmod\ x^2 + 1)$ is an equivalence relation on $\mathbb{R}[x]$.

9.6 Check that addition and multiplication in Definition 9.6 do not depend on representatives chosen from the equivalence class.

That is, suppose we have polynomials $p_1(x)$, $p_2(x)$, $q_1(x)$, and $q_2(x)$ with

$$p_1(x) \equiv p_2(x) \quad \text{and} \quad q_1(x) \equiv q_2(x).$$

Show that

$$p_1(x) + q_1(x) \equiv p_2(x) + q_2(x) \quad \text{and} \quad p_1(x) \cdot q_1(x) \equiv p_2(x) \cdot q_2(x).$$

9.7 Let us say that polar coordinates are *equivalent* if they define the same complex number.

Determine conditions on r_1, r_2, θ_1, and θ_2 for when (r_1, θ_1) and (r_2, θ_2) are equivalent that do not use trigonometric functions.

9.8 Let $z = a + bi$ be a complex number and define the matrix M_z by

$$M_z = \begin{bmatrix} a & b \\ -b & a \end{bmatrix}.$$

a) Show that $M_{w+z} = M_w + M_z$.

b) Show that $M_{wz} = M_w \cdot M_z$.

c) What can you say about $\det M_z$?

9.9 The complex numbers are formed from the real numbers by appending a new number, i, and then following the usual rules of algebra. In a similar way, we can append i to the modular numbers $\mathbb{Z}_3$ to create a new field $\mathbb{F}_9$. The elements of $\mathbb{F}_9$ are of the from $a + bi$ where $a, b \in \mathbb{Z}_3$.

a) Write down the addition and multiplication tables for $\mathbb{F}_9$.

b) Check that every nonzero element of $\mathbb{F}_9$ has a multiplicative inverse.

c) The nonzero elements of $\mathbb{F}_9$ are denoted $\mathbb{F}_9^*$. We showed[1] that $(\mathbb{Z}_7^*, \cdot)$ and $(\mathbb{Z}_6, +)$ are isomorphic. Find an isomorphism between $(\mathbb{F}_9^*, \cdot)$ and $(\mathbb{Z}_8, +)$.

d) Appending i to $\mathbb{Z}_3$ yields a field. Show that's not the case for appending i to $\mathbb{Z}_5$.

Addendum: Complex Exponential

The complex number with polar coordinates (r, θ) is converted to the standard representation $a + bi$ by these formulas:

$$a = r \cos \theta \quad \text{and} \quad b = r \sin \theta.$$

There is an alternative conversion using Euler's number, e. We do this by extending the exponential function e^x to complex exponents.

Proposition 9.10. *For a real number θ we have* $e^{i\theta} = \cos \theta + i \sin \theta$.

This implies $e^{a+bi} = e^a e^{bi} = e^a (\cos b + i \sin b)$.

A formal proof of Proposition 9.10 is beyond the scope of this book, but we can show why this is true using the power series representations for e^x, $\sin x$,

---
[1] See the discussion beginning on page 125.

and $\cos x$. There are those formulas:

$$e^x = \frac{x^0}{0!} + \frac{x^1}{1!} + \frac{x^2}{2!} + \frac{x^3}{3!} + \cdots,$$

$$\sin x = \frac{x^1}{1!} - \frac{x^3}{3!} + \frac{x^5}{5!} - \frac{x^7}{7!} + \cdots,$$

$$\cos x = \frac{x^0}{0!} - \frac{x^2}{2!} + \frac{x^4}{4!} - \frac{x^6}{6!} + \cdots.$$

We substitute $i\theta$ for x in the formula for e^x and collect the real and imaginary parts:

$$\begin{aligned} e^{i\theta} &= \frac{(i\theta)^0}{0!} + \frac{(i\theta)^1}{1!} + \frac{(i\theta)^2}{2!} + \frac{(i\theta)^3}{3!} + \frac{(i\theta)^4}{4!} + \frac{(i\theta)^5}{5!} + \frac{(i\theta)^6}{6!} + \frac{(i\theta)^7}{7!} + \cdots \\ &= \frac{\theta^0}{0!} + \frac{\theta^1}{1!}i - \frac{\theta^2}{2!} - \frac{\theta^3}{3!}i + \frac{\theta^4}{4!} + \frac{\theta^5}{5!}i - \frac{\theta^6}{6!} - \frac{\theta^7}{7!}i + \cdots \\ &= \left(\frac{\theta^0}{0!} - \frac{\theta^2}{2!} + \frac{\theta^4}{4!} - \frac{\theta^6}{6!} + \cdots \right) + \left(\frac{\theta^1}{1!} - \frac{\theta^3}{3!} + \frac{\theta^5}{5!} - \frac{\theta^7}{7!} + \cdots \right) i \\ &= \cos\theta + i\sin\theta. \end{aligned}$$

The consequence of Proposition 9.10 is that the complex number with polar coordinates (r, θ) is $re^{i\theta}$.

From this we see that when we multiply the complex numbers with polar coordinates (r, θ) and (s, ϕ) we have

$$\begin{aligned} (r, \theta) \cdot (s, \phi) &= \left(re^{i\theta}\right) \cdot \left(se^{i\phi}\right) \\ &= rse^{i\theta}e^{i\phi} \\ &= rse^{i(\theta+\phi)} \\ &= (rs, \theta + \phi), \end{aligned}$$

reaffirming Proposition 9.8.

Chapter 10

Further Extensions

There is a natural progression from counting to the complex numbers. Each step along this journey from $\mathbb{N}$ to $\mathbb{C}$ repairs "defects;" that is, in each new number system we are able to solve equations that had no solution in its predecessor. Delightfully, when we arrive at $\mathbb{C}$ we find that all polynomial equations are solvable; the complex numbers are algebraically closed.

Where to next? Are there other number systems to explore? Indeed there are many, but none is definitively "the next step" after $\mathbb{C}$. In this chapter we give a brief sampling of some of these extensions just to whet your appetite for some exotic ideas.

This chapter is a teaser and gives only a lightning introduction to a handful of other number concepts.

10.1 Infinities

The real and complex numbers are all *finite*; there is no number called "infinity" in either $\mathbb{R}$ or $\mathbb{C}$. In this section we present a few ways in which ∞ takes its place along with other numbers.

Real Numbers and $\pm\infty$

The *extended* real numbers contain all the real numbers and two additional elements: ∞ and $-\infty$. The set of extended real numbers is denoted $\overline{\mathbb{R}} = \mathbb{R} \cup \{-\infty, \infty\}$.

When we incorporate these two new members, we also need to specify their behavior with the operations of addition and multiplication, and we need to extend the definition of $\leq$.

Some decisions are simple. For example, if a is an (ordinary) real number then we have the following:

$$a + \infty = \infty, \qquad a + (-\infty) = -\infty,$$
$$\infty + \infty = \infty, \qquad (-\infty) + (-\infty) = -\infty.$$

However, in $\overline{\mathbb{R}}$ the expression $\infty + (-\infty)$ is undefined.

Multiplication is only slightly more complicated. For a positive (ordinary) real number a we have these definitions:

$$a \cdot \infty = \infty, \qquad (-a) \cdot \infty = -\infty,$$
$$a \cdot (-\infty) = -\infty, \qquad (-a) \cdot (-\infty) = \infty,$$

and these:

$$\infty \cdot \infty = \infty, \qquad (-\infty) \cdot (-\infty) = \infty, \qquad \infty \cdot (-\infty) = -\infty.$$

However, $\infty \cdot 0$ is undefined.

The upshot is that neither ∞ nor $-\infty$ has additive and multiplicative inverses, but we do have this for any real number a:

$$\frac{a}{\infty} = \frac{a}{-\infty} = 0.$$

However, ∞/∞ is undefined.

The number system $\overline{\mathbb{R}}$ is not a field.

Extending $\leq$ is straightforward. For any real number a we have $-\infty < a < \infty$.

Complex Infinity

We may also extend the complex numbers, but in this case we do so with a single *complex* infinity, which we denote as ∞. The extended complex numbers are denoted $\overline{\mathbb{C}} = \mathbb{C} \cup \{\infty\}$.

Addition is rather simple: For any complex number z we have $z + \infty = \infty$ and, of course, $\infty + \infty = \infty$.

Multiplication by nonzero values is also simple: If $z \neq 0$ then $\infty \cdot z = \infty$ (and $\infty \cdot \infty = \infty$). However, $0 \cdot \infty$ is undefined.

In the extended complex numbers $-\infty$ is the same as ∞, but $\infty - \infty$ is not defined.

Division by ∞ always gives 0 as the result except that ∞/∞ is undefined.

Tropical Arithmetic

The tropical number system appends ∞ to the real numbers, and provides alternatives for addition and multiplication. To distinguish these new operations from the standard ones, we use the symbols $\oplus$ and $\odot$.

> **Computing with ∞**
>
> Some computer programming languages allow the values $+\infty$ and $-\infty$ in calculations. For example, in the Julia language the result of $1/0$ is $+\infty$ and $-1/0$ is $-\infty$. However, attempts to perform undefined calculations return the value NaN which stands for not-a-number.
>
> ```
> julia> 1/0
> Inf
>
> julia> -1/0
> -Inf
>
> julia> 1/Inf
> 0.0
>
> julia> Inf+Inf
> Inf
>
> julia> Inf-Inf
> NaN
>
> julia> Inf*Inf
> Inf
>
> julia> Inf/Inf
> NaN
> ```

The $\oplus$ sum of two numbers is the smaller[1] of the pair; in notation:

$$a \oplus b = \min\{a, b\}.$$

For example, $9 \oplus 3 = 3$ because 3 is the smaller of 3 and 9.

The identity element for $\oplus$ is ∞ because

$$a \oplus \infty = \min\{a, \infty\} = a.$$

Notice, however, that real numbers do not have tropical additive inverses.

Tropical multiplication of a and b is the ordinary sum of a and b:

$$a \odot b = a + b.$$

For example $3 \odot 5 = 8$. Notice that 1 is *not* the identity element for tropical multiplication; zero is!

$$a \odot 0 = a + 0 = a.$$

---

[1] There is an equivalent version of tropical arithmetic in which $a \oplus b$ is the larger of a and b, in which case $-\infty$ is used instead of $+\infty$.

All (finite) real numbers have tropical multiplicative inverses because

$$a \odot (-a) = a + (-a) = 0,$$

but ∞ does not have a multiplicative inverse.

Interestingly, the distributive property holds. For any $a, b, c \in \mathbb{R} \cup \{\infty\}$ we have

$$\begin{aligned} a \odot (b \oplus c) &= a \odot \min\{b, c\} \\ &= \min\{a + b, a + c\} \\ &= \min\{a \odot b, a \odot c\} \\ &= (a \odot b) \oplus (a \odot c). \end{aligned}$$

Hyperreal Numbers

The hyperreal numbers are a mammoth extension to the reals, $\mathbb{R}$. Not only do they incorporate infinitely large values, they also include infinitesimals – positive numbers that are smaller than any positive real number. The hyperreal numbers, denoted $^*\mathbb{R}$, are the invention of the twentieth-century mathematician Abraham Robinson.

In broad strokes, the hyperreal numbers include a number s with the property that $0 < s < 1/N$ for all positive integers N. Notice then that $1/s$ is a number that is greater than all integers.

Hyperreal numbers can be added, multiplied, and ordered by the usual less-than-or-equal relation. Indeed $(^*\mathbb{R}, +, \cdot, \leq)$ forms an ordered field.

It takes a Herculean effort to define $^*\mathbb{R}$ properly, but with that effort in place intuitive ideas from calculus become rather natural. For example, consider the function $f(x) = x^2$. The standard definition of the derivative of f is this:

$$f'(x) = \lim_{h \to 0} \frac{f(x+h) - f(x)}{h}.$$

Working in $^*\mathbb{R}$ we can avoid any mention of limit. Using an infinitesimal number s, inside $^*\mathbb{R}$ we calculate:

$$\begin{aligned} \frac{f(x+s) - f(x)}{x} &= \frac{(x+s)^2 - x^2}{s} \\ &= \frac{x^2 + 2sx + s^2 - x^2}{s} \\ &= \frac{2sx + s^2}{s} \\ &= 2x + s. \end{aligned}$$

Note that there was no division by zero.

Finally, the derivative $f'(x)$ is defined to be the "real part" of $2x + s$; in other words, $2x$ is an ordinary real number that is infinitesimally close to $2x + s$. The theory of hyperreal numbers allows us, with full logical rigor, to ignore the $+s$ term and conclude that $f'(x) = 2x$.

Transfinite Cardinals

The natural numbers are defined as the sizes of *finite* sets. The nineteenth-century German mathematician Georg Cantor introduced a new suite of numbers to represent the sizes of *infinite* sets.

At first glance this seems simple; just use the symbol ∞ to stand for the size of an infinite set. However, recall that finite sets have the same size exactly when there is a one-to-one correspondence between the two sets. Cantor showed that there are pairs of infinite sets that do not have such a matching.

The simplest infinite set is the natural numbers: $\mathbb{N} = \{0, 1, 2, \ldots\}$. The *power set* of $\mathbb{N}$ is the set of all subsets of $\mathbb{N}$. The notation for the power set of $\mathbb{N}$ is $2^{\mathbb{N}}$:
$$2^{\mathbb{N}} = \{A : A \subseteq \mathbb{N}\}.$$
Clearly $2^{\mathbb{N}}$ is also an infinite set. Let's see that there is no one-to-one correspondence between $\mathbb{N}$ and $2^{\mathbb{N}}$.

Suppose (for the sake of contradiction) that there is a one-to-one correspondence between $\mathbb{N}$ and $2^{\mathbb{N}}$. That means we have a pairing that looks like this:
$$0 \leftrightarrow A_0, \quad 1 \leftrightarrow A_1, \quad 2 \leftrightarrow A_2, \quad 3 \leftrightarrow A_3, \quad 4 \leftrightarrow A_4, \quad \cdots$$
where the sets on the right of the arrows are all of the subsets of $\mathbb{N}$. Now consider this set:
$$B = \{n \in \mathbb{N} : n \notin A_n\}.$$
In words: B is a subset of $\mathbb{N}$ in which $n \in B$ exactly when $n \notin A_n$. For example, if A_0 includes 0, then B does not; conversely if $0 \notin A_0$ then $0 \in B$.

Since B is a subset of $\mathbb{N}$ it is somewhere on the right side of the list, say $k \leftrightarrow A_k = B$.

Is $k \in B$? If it is, then $k \in A_k$ and so, by the definition of B, we conclude that $k \notin B$.

OK, so $k \notin B$. That means $k \notin A_k$ which, by the definition of B, means that $k \in B$.

In other words, $k \in B$ implies $k \notin B$, and $k \notin B$ implies $k \in B$! That's clearly impossible.

Therefore, there is no one-to-one correspondence between $\mathbb{N}$ and $2^{\mathbb{N}}$.

Furthermore, clearly $2^{\mathbb{N}}$ is a larger set than $\mathbb{N}$ because there is a one-to-one correspondence between $\mathbb{N}$ and a *subset* of $2^{\mathbb{N}}$ like this:
$$0 \leftrightarrow \{0\} \quad 1 \leftrightarrow \{1\} \quad 2 \leftrightarrow \{2\} \quad 3 \leftrightarrow \{3\} \quad 4 \leftrightarrow \{4\} \quad \cdots$$

Cantor used the notation[2] $\aleph_0$ to stand for the *cardinality* (i.e., the size) of the set $\mathbb{N}$. This argument shows that the size of $2^{\mathbb{N}}$ is greater than $\aleph_0$.

By a similar argument, we can show that the cardinality of $\mathbb{R}$ is also greater than $\aleph_0$. In fact, there is a one-to-one correspondence between $\mathbb{R}$ and $2^{\mathbb{N}}$.

---

[2] The notation $\aleph_0$ is pronounced *aleph-null* or *aleph-naught*. Aleph is the first letter of the Hebrew alphabet.

Transfinite Ordinals

Natural numbers $\mathbb{N}$ and transfinite cardinal numbers such as $\aleph_0$ are used to describe the size – the cardinality – of sets. These numbers arise from one-to-one correspondences of sets. For example, the natural number 3 is the equivalence class of all sets that have a one-to-one correspondence with $\{a, b, c\}$ (where a, b, and c are distinct objects). The transfinite number $\aleph_0$ is the equivalence class of all sets that have a one-to-one correspondence with $\mathbb{N}$.

As discussed in Section 1.1, sets are *unordered*; there is no difference between $\{1, 2, 3\}$ and $\{2, 1, 3\}$. However, some sets, such as $\mathbb{N}$ do have an order associated with their elements: $0 < 1 < 2 < 3 < \cdots$. When a set has an ordering for its elements, we call it an ordered set and we show this by writing $\{1 < 2 < 3\}$.

Suppose A and B are ordered sets. We say they are *order equivalent* if there is a one-to-one correspondence between the sets that respects their orderings. This means that if a_1 and a_2 are elements of A, with $a_1 < a_2$, and these are paired with elements of B like this $a_1 \leftrightarrow b_1$ and $a_2 \leftrightarrow b_2$, then we require $b_1 < b_2$.

For example, the ordered sets $A = \{1 < 2 < 3\}$ and $B = \{6 < 7 < 8\}$ are order equivalent because we can pair

$$1 \leftrightarrow 6, \quad 2 \leftrightarrow 7, \quad \text{and} \quad 3 \leftrightarrow 8.$$

There are other one-to-one correspondences between these two sets, but they are not order preserving.

For ordered sets A and B, let $A \equiv B$ mean that there is an order-preserving one-to-one correspondence between A and B.

Recall that the natural numbers enjoy the well-ordering property (see page 40): every nonempty subset of $\mathbb{N}$ contains a least element.

The well-ordering property also holds for any subset of $\mathbb{N}$, but not for other ordered sets such as $\mathbb{Q}$ or $\mathbb{R}$.

An *ordinal number* is an $\equiv$-equivalence class of well-ordered sets. For example, $4 = [\![\{a < b < c < d\}]\!]$.

Addition of ordinal numbers is accomplished by taking (disjoint) representatives and placing one after the other. For example, given

$$4 = [\![\{a < b < c < d\}]\!] \quad \text{and} \quad 3 = [\![\{x < y < z\}]\!],$$

we have

$$3 + 4 = [\![\{a < b < c < d < x < y < z\}]\!]$$

and

$$4 + 3 = [\![\{x < y < z < a < b < c < d\}]\!].$$

Since there is an order-preserving one-to-one correspondence between the ordered sets $\{a < b < c < d < x < y < z\}$ and $\{x < y < z < a < b < c < d\}$ we have the unexciting conclusion that $3 + 4 = 4 + 3$.

In the case of finite sets, there is no difference between ordinal and cardinal numbers. Things get interesting when we look at infinite sets.

Let's consider two ways to extend $\mathbb{N}$ with a single additional element. First, we can insert a new element – let's call it a – that is less than all other natural numbers and make a new ordered set A:

$$A = \{a < 0 < 1 < 2 < 3 < \cdots\}.$$

Note that there is an order-preserving one-to-one correspondence between $\mathbb{N}$ and this new set A:

$$\begin{array}{cccccccc} \mathbb{N}: & 0 & 1 & 2 & 3 & 4 & \cdots \\ & \updownarrow & \updownarrow & \updownarrow & \updownarrow & \updownarrow & \\ A: & a & 0 & 1 & 2 & 3 & \cdots \end{array}$$

In symbols, $A \equiv \mathbb{N}$. Let ω be the ordinal number $\omega = [\![\mathbb{N}]\!] = [\![A]\!]$. Notice that prepending a single element to $\mathbb{N}$ yields an equivalent set giving us the equation $1 + \omega = \omega$. Intuitively, this makes some sense; adding one thing to an infinite set results in an infinite set.

Another way to extend $\mathbb{N}$ is to add a new element that is greater than all elements of $\mathbb{N}$. Let's call that new element z and create the ordered set Z as

$$Z = \{0 < 1 < 2 < 3 < \cdots < z\}.$$

In this case $Z \not\equiv \mathbb{N}$ because we can't pair the new element z to anything in $\mathbb{N}$ and have order preserved. This gives the interesting equation $\omega + 1 \neq \omega$.

The upshot is that addition of ordinal numbers is not commutative, and that is the *least* strange thing about transfinite ordinal numbers!

10.2 Quaternions

Complex numbers are created from real numbers by the addition of a new number named i with the property that $i^2 = -1$. The usual arithmetic operations of addition and subtraction give rise to the entire set $\mathbb{C}$, whose elements are of the form $a + bi$ where $a, b \in \mathbb{R}$. Taking $a = 0$ and $b = -1$ we see there is a second square root of -1, namely $-i$.

The nineteenth-century Irish mathematician William Rowan Hamilton created an extension to the complex numbers called the *quaternions*.

Hamilton's idea was to append to $\mathbb{R}$ three distinct new numbers called i, j, and k. They (and their negatives) are all square roots of -1:

$$i^2 = j^2 = k^2 = -1.$$

A *quaternion* is a number of the form $a + bi + cj + dk$ where $a, b, c, d \in \mathbb{R}$. The set of all quaternions is denoted $\mathbb{H}$ in Hamilton's honor.

Addition of quaternions is as one might expect:

$$(a + bi + cj + dk) + (w + xi + yj + zk) = (a + w) + (b + x)i + (c + y)j + (d + z)k.$$

Multiplication, however, is exciting. To begin, we need to define what happens when i, j, and k are multiplied with each other. Hamilton's rules for this are as follows:

$$ij = k, \quad jk = i, \quad ki = j,$$
$$ji = -k, \quad kj = -i, \quad ik = -j.$$

Notice that multiplication is *not* commutative. The properties of i, j, and k can be compactly summarized in a single line[3] like this:

$$i^2 = j^2 = k^2 = ijk = -1.$$

Using the basic properties of i, j, and k, and the usual rules of algebra, we have this:

$$\begin{aligned}(a + bi + cj + dk) \cdot (w + xi + yj + zk) = &(aw - bx - cy - dz) \\ &+ (ax + bw + cz - dy)i \\ &+ (ay + cw - bz + dx)j \\ &+ (az + dw + by - cx)k.\end{aligned}$$

It is not hard to see that $0 + 0i + 0j + 0k$ is the identity element for addition and that all members of $\mathbb{H}$ have additive inverses. The identity element for multiplication is $1 + 0i + 0j + 0k$, and one can check that the multiplicative inverse of $a + bi + cj + dk$ is

$$\frac{a}{T} - \frac{b}{T}i - \frac{c}{T}j - \frac{d}{T}k$$

where $T = a^2 + b^2 + c^2 + d^2$. This implies that all nonzero elements of $\mathbb{H}$ have multiplicative inverses.

In all, $(\mathbb{H}, +, \cdot)$ satisfies all the conditions of being a field, except that multiplication is not commutative. Such an algebraic structure is known as a *skew field*.

10.3 *p*-adic Numbers

A Quick Review of Place-value Notation

Real numbers are often expressed in decimal notation. In this notation we begin with an optional minus sign (for negative numbers), a finite list of digits (the

---

[3] Hamilton, excited by his insight into how to define quaternions, carved this single-line equation into a stone on a bridge in Dublin. A plaque now commemorates this bit of mathematical graffiti. See [15].

10 Further Extensions   161

symbols 0 through 9), a decimal point, and then an infinite list of digits. This is a place-value system and the contribution of a digit depends on where it is situated in the notation.

For example, in the number 23.9078... the 2 contributes 2×10^1 to the value and the 7 contributes 7×10^{-3}.

Base-five notation (see Exercise 2.27) is an alternative way to express real numbers. The format is the same as decimal notation, but we use only the digits 0 through 4. As in decimal notation the location of a digit determines its contribution to the value.

Consider the number 34.1_{FIVE}. The 3 contributes 3×5^1, the 4 contributes 4×5^0, and the 1 contributes 1×5^{-1}. Therefore

$$34.1_{\text{FIVE}} = 3 \times 5^1 + 4 \times 5^0 + 1 \times 5^{-1} = 15 + 4 + 0.2 = 19.2_{\text{TEN}}$$

Whether in base ten or base five, the digits further and further to the right of the radix point[4] have less and less significance. For example, in base ten we can see that the difference between π and $355/113$ is small

$$\pi = 3.141592\underline{6}53589793\ldots$$

$$\frac{355}{113} = 3.141592\underline{9}20353982\ldots$$

because these two numbers agree to six decimal places.

Salient points:

- There are finitely many digits to the left of the radix point and infinitely many to the right.

- If the first position where two numbers disagree is far to the right, then the two numbers are close together.

We're about to turn all this on its head.

Nonsense

We know that $0.9999\ldots$ is equal to 1. In base five, we have the analogous $0.4444\ldots_{\text{FIVE}} = 1$. Let's review why that's true.

The expression $0.4444\ldots_{\text{FIVE}}$ can be written like this:

$$0.4444\ldots_{\text{FIVE}} = 4 \cdot 5^{-1} + 4 \cdot 5^{-2} + 4 \cdot 5^{-3} + 4 \cdot 5^{-4} + \cdots.$$

Call that number x. Here is $\frac{1}{5}x$:

$$\frac{1}{5}x = \frac{1}{5} \cdot \left(4 \cdot 5^{-1} + 4 \cdot 5^{-2} + 4 \cdot 5^{-3} + 4 \cdot 5^{-4} + \cdots\right)$$
$$= 4 \cdot 5^{-2} + 4 \cdot 5^{-3} + 4 \cdot 5^{-4} + 4 \cdot 5^{-5} + \cdots.$$

---

[4] In base ten, this is called the decimal point; but since we are working in other bases, the dot is called a *radix point*.

Next, subtract $\frac{1}{5}x$ from x:

$$x = 4 \cdot 5^{-1} + 4 \cdot 5^{-2} + 4 \cdot 5^{-3} + 4 \cdot 5^{-4} + \cdots$$
$$- \quad \tfrac{1}{5}x = \qquad\qquad 4 \cdot 5^{-2} + 4 \cdot 5^{-3} + 4 \cdot 5^{-4} + \cdots$$
$$\overline{\tfrac{4}{5}x = 4 \cdot 5^{-1}.}$$

Therefore $x = 1$.

Let's do this again, but this time let

$$x = 4 + 4 \cdot 5 + 4 \cdot 5^2 + 4 \cdot 5^3 + 4 \cdot 5^4 + \cdots$$

Were we to write this in base five, we would have this: $\ldots 4444_{\text{FIVE}}$ with infinitely many 4s to the left. This is ridiculous, but stay with it.

Let's calculate $5x$:

$$5x = 5\left(4 + 4 \cdot 5 + 4 \cdot 5^2 + 4 \cdot 5^3 + 4 \cdot 5^4 + \cdots\right)$$
$$= 4 \cdot 5 + 4 \cdot 5^2 + 4 \cdot 5^3 + 4 \cdot 5^4 + 4 \cdot 5^5 + \cdots.$$

In base five, we'd write this as $\ldots 444440_{\text{FIVE}}$.

Following the ideas from before, let's evaluate $x - 5x$:

$$x = 4 + 4 \cdot 5 + 4 \cdot 5^2 + 4 \cdot 5^3 + 4 \cdot 5^4 + 4 \cdot 5^5 + \cdots$$
$$- \quad 5x = \qquad 4 \cdot 5 + 4 \cdot 5^2 + 4 \cdot 5^3 + 4 \cdot 5^4 + 4 \cdot 5^5 + \cdots$$
$$\overline{-4x = 4}$$

and therefore $x = -1$.

Just to "check" this is correct, add 1 to $\ldots 44444_{\text{FIVE}}$ using the usual rules of addition. That is, we add 1 to the rightmost 4, giving 0 (because we're in base five) and carry the 1. Add that 1 to the next-to-last 4 to give another 0 and carry 1 again. Keep going and we have this:

$$\begin{array}{r} \ldots 444444_{\text{FIVE}} \\ + \qquad\qquad 1_{\text{FIVE}} \\ \hline \ldots 000000_{\text{FIVE}} \end{array}$$

This shows that $\ldots 44444_{\text{FIVE}} = -1$.

We should also expect that squaring $\ldots 44444_{\text{FIVE}}$ to yield 1. Let's do the calculation. Multiplication proceeds right-to-left with the usual carries; just remember that we are working in base five so when we multiply the rightmost 4s together the result is $16_{\text{TEN}} = 31_{\text{FIVE}}$. Therefore we record a 1 and carry a 3 to the next column. The calculation looks like this (with the FIVE subscripts omitted

10 Further Extensions

for clarity):

$$
\begin{array}{r}
\ldots 4444444 \\
\times \quad \ldots 4444444 \\
\hline
\ldots 444444441 \\
\ldots 44444441 \\
\ldots 4444441 \\
\ldots 444441 \\
+ \quad \ldots 44441 \\
\hline
\ldots 000000001
\end{array}
$$

Thus $(\ldots 4444444_{\text{FIVE}})^2 = 1$ consistent with our assertion that $\ldots 4444444_{\text{FIVE}} = -1$.

Just for fun, let's evaluate $x = \ldots 22222223_{\text{FIVE}}$. We multiply this by 5 to get the digits to line up and subtract:

$$
\begin{array}{r}
5x = \ldots 22222230_{\text{FIVE}} \\
- \quad x = \ldots 22222223_{\text{FIVE}} \\
\hline
4x = \ldots 00000002_{\text{FIVE}}
\end{array}
$$

giving $4x = 2$ and so $x = \frac{1}{2}$.

To check this, multiply $\ldots 2222223_{\text{FIVE}}$ by 2 in base five:

$$
\begin{array}{r}
\ldots 222222223_{\text{FIVE}} \\
\times \quad 2_{\text{FIVE}} \\
\hline
\ldots 000000001_{\text{FIVE}}
\end{array}
$$

because the initial 2×3 gives 1 with a carry of 1. Then 2×2 plus the carry of 1 gives 0 with a carry of 1, and then we cascade out to the left giving the infinite stream of 0s.

Given that $\ldots 22222223_{\text{FIVE}}$ is one-half, if we divide this by 5 we get one-tenth. In base five, division by 5 is easy: just slide all the digits one place to the right, so this should give:

$$\frac{1}{10} = \ldots 2222222.3_{\text{FIVE}}.$$

As an independent confirmation, we know that $\ldots 444444_{\text{FIVE}}$ equals -1, and so $\ldots 222222_{\text{FIVE}}$ is $\frac{1}{2}$. Add to that $\ldots 000000.3_{\text{FIVE}} = \frac{3}{5}$ and we find

$$\ldots 222222.3_{\text{FIVE}} = \ldots 222222_{\text{FIVE}} + \ldots 000000.3_{\text{FIVE}} = -\frac{1}{2} + \frac{3}{5} = \frac{1}{10}.$$

Making Sense

We have made the ridiculous claim that ...$444444_{\text{FIVE}} = -1$. Written out in full in base ten, we're asserting that

$$4 + 20 + 100 + 500 + 250 + 12{,}500 + \cdots = -1.$$

And yet all the calculations are consistent. Perhaps this is not utter nonsense.

Recall from Chapter 7 (see page 107) that ordinary decimal numbers, such as 3.14159... are sequences:

$$\hat{p} = (3, 3.1, 3.14, 3.141, 3.1415, 3.14159, \ldots).$$

What's more, these are Cauchy sequences and therefore represent real numbers.

Recall that in a Cauchy sequence, the terms gets closer and closer together the further we progress into the sequence. When written as decimals, two rational numbers are close when their first discrepancy is far to the right of the decimal point.

Inverting this, let's consider two numbers with infinitely many digits before the radix point to be close if the first discrepancy is far to the left. In this way these

$$a = \ldots 240113\underline{1}22230014.1135_{\text{FIVE}}$$
$$b = \ldots 114320\underline{4}22230014.1135_{\text{FIVE}}$$

are close together but these

$$c = \ldots 334011122340.114\underline{3}1114_{\text{FIVE}}$$
$$d = \ldots 334011122340.114\underline{0}2322_{\text{FIVE}}$$

are far apart.

The discrepancy between a and b is 9 places to the left of the radix point; this is the 5^8-column. The difference between c and d is 4 places to the right of the radix point; that is the 5^{-4} column. Ordinarily we'd say the distance between a and b is (roughly) 5^8 and the difference between c and d is (roughly) 5^{-4}. However, since we are doing things backwards, we invent a new notion of closeness by creating a "distance" that says a and b are 5^{-8} apart while c and d are 5^4 apart. We accomplish this by replacing ordinary absolute value $|x|$ with a "backward" version denoted $|x|_5$ that we develop now.

Every positive rational r number is a fraction x/y where x and y are positive integers. Either of x or y might be a multiple of five.[5] We can pull that factor of 5 out in front and write r like this:

$$r = 5^t \cdot \frac{a}{b}$$

---

[5] There is nothing special about the number 5 in this running example. In general we can do this work in any number base, although prime bases are preferred as explained later.

where neither a nor b is a multiple of 5. For example, if $r = 20/3$ we write that as
$$r = \frac{20}{3} = 5^1 \cdot \frac{4}{3}.$$
Likewise, if $r = 7/100$ then rewrite it as
$$r = \frac{7}{100} = 5^{-2} \cdot \frac{7}{4}.$$

We now define a replacement for absolute value. If $r = 5^t \cdot \frac{a}{b}$ (where neither a nor b is divisible by 5) we have
$$|r|_5 = \left|5^t \cdot \frac{a}{b}\right|_5 = 5^{-t}.$$

Notice that the larger t is, the smaller $|r|_5$ is. Let's examine a few examples:

$$\left|\frac{20}{3}\right|_5 = \left|5^1 \cdot \frac{4}{3}\right|_5 = 5^{-1}, \qquad \left|\frac{7}{1000}\right|_5 = \left|5^{-3}\frac{7}{4}\right|_5 = 5^3,$$

$$\left|\frac{4}{3}\right|_5 = \left|5^0 \cdot \frac{4}{3}\right|_5 = 5^0, \qquad |6000|_5 = \left|5^3 \cdot \frac{48}{1}\right|_5 = 5^{-3}.$$

Going forward, we assess the closeness of rational numbers a and b using $|a - b|_5$ instead of $|a - b|$.

In the context of the real number system, the expression $0.999999\ldots$ is, in fact, a Cauchy sequence:
$$(0.9, 0.99, 0.999, 0.9999, \ldots)$$
whose terms get closer and closer to 1. Indeed, the nth term is $1 - 10^{-n}$, so the difference between the nth term and 1 is 10^{-n}. The sequence converges to 1 and that's why $0.9999\ldots$ is equal to 1.

Let's switch our context to the 5-adic number $\ldots 444444_{\text{FIVE}}$. This is also a sequence:
$$\hat{a} = (4_{\text{FIVE}}, 44_{\text{FIVE}}, 444_{\text{FIVE}}, 4444_{\text{FIVE}}, 44444_{\text{FIVE}}, \ldots).$$

The nth term of this sequence is
$$a_n = \underbrace{444\ldots 4}_{n}{}_{\text{FIVE}} = 4 + 4 \cdot 5 + 4 \cdot 5^2 + \cdots 4 \cdot 5^{n-1}$$
$$= 4\left[1 + 5 + 5^2 + \cdots + 5^{n-1}\right]$$
$$= 4\left[\frac{1 - 5^n}{1 - 5}\right]$$
$$= 5^n - 1.$$

The 5-adic distance between a_n and -1 is therefore

$$|a_n - (-1)|_5 = |(5^n - 1) + 1|_5 = 5^{-n}$$

which gets smaller and smaller as n gets larger. Thus the terms in $\hat{a}$ converge to -1 justifying the "nonsensical"

$$\ldots 4444444_{\text{FIVE}} = -1.$$

The 5-adic Numbers, $\mathbb{Q}_5$

Real numbers can be written, in base ten, optionally starting with a minus sign and then followed by a finite list of digits, a decimal point, and then an infinite stream of digits to the right. In this notation, rational numbers eventually start repeating sequences of digits. For example,

$$\frac{431}{330} = 1.30606060606060\ldots$$

We have been exploring numbers of the form $\ldots 40432401.44_{\text{FIVE}}$. Here, there is no sign and there are infinitely many digits to the left of the radix point but only finitely many to the right. Such numbers are known as 5-adic numbers.

They can correctly be interpreted as sequences. For example, the 5-adic number $\ldots 40432401.44_{\text{FIVE}}$ is this sequence:

$$(0.04_{\text{FIVE}}, 0.44_{\text{FIVE}}, 1.44_{\text{FIVE}}, 1.44_{\text{FIVE}}, 401.44_{\text{FIVE}}, 2401.44_{\text{FIVE}}, 32401.44_{\text{FIVE}}, \ldots).$$

Using $|a - b|_5$ as measure of closeness, these sequences are truly Cauchy sequences.

Just as rational numbers have repeating decimals in ordinary notation, in 5-adic notation rational numbers develop repeating patterns when we read them from right to left.

The full collection of 5-adic numbers (including those with repeating patterns and those without) is denoted $\mathbb{Q}_5$. The rational numbers form a proper subset of $\mathbb{Q}_5$, but the full set $\mathbb{Q}_5$ is not the same as the real numbers. It is an entirely new set of numbers altogether.

The 5-adic numbers can be added, subtracted, and multiplied. It turns out that nonzero 5-adic numbers have multiplicative inverses. In other words, $\mathbb{Q}_5$ is a field that contains the rational numbers, but is not isomorphic to $\mathbb{R}$.

How do we know $\mathbb{Q}_5$ is not the same as $\mathbb{R}$, but with strange names for its elements? Much of our earlier effort was to show that $\mathbb{R}$ contains a square root of 2. We note that $\mathbb{Q}_5$ does not! Here's why:

Suppose there were an element $a \in \mathbb{Q}_5$ with $a^2 = 2$. We consider two possible cases: a has some nonzero digits to the right of its radix point or not.

In the first case a has a form that looks like this:

$$a = \ldots \square\square\square\square\square\square\square\square.\square\square\square\square X_{\text{FIVE}}$$

where X is a nonzero digit; thus X is one of 1, 2, 3, or 4. When we square a, that last digit multiplies by itself and produces the new rightmost digit. Note that since X is one of 1, 2, 3, or 4, then that digit multiplied by itself it yields either a 1 (in case X is 1 or 4) or a 4 (in case X is 2 or 3). In other words, in a^2 there are nonzero digits to the right of the radix point and so $a^2 \neq 2$.

The other possibility is that a has no nonzero digits to the right of the radix point. In that case a looks like this:

$$a = \ldots \square\square\square\square\square\square\square\square\square Y_{\text{FIVE}}$$

where Y is one of 0, 1, 2, 3, or 4. However, when we multiply a by itself the rightmost digit is one of 0 (if $Y = 0$), 1 (if Y is 1 or 4), or 4 (if Y is 2 or 3). However, $a^2 = \ldots 00000002_{\text{FIVE}}$ and in no case does a 2 appear in the ones column.

Therefore there is no square root of 2 in $\mathbb{Q}_5$ but there is a $\sqrt{-1}$ in $\mathbb{Q}_5$; see Exercise 10.9.

Recap

There is a natural progression from simple counting to the real numbers, and from the real numbers to the complex numbers. We saw in this chapter that $\mathbb{C}$ need not be the end of our journey; there are now many paths forward. In this chapter we explored some exotic extensions to the notion of number including various ways to think about infinity, additional square roots of -1 in the quaternions, and the new worlds of p-adic numbers. This is not an exhaustive list; there are many other extensions, some of which are truly surreal! (See [5].)

Exercises

10.1 A common mistake for new algebra students is to "simplify" the expression $(x + y)^2$ as $x^2 + y^2$.

Show that in tropical arithmetic $(x \oplus y) \odot (x \oplus y)$ equals $(x \odot x) \oplus (y \odot y)$.

10.2 Our development of various number systems is motivated, in part, by unsolvable equations. For example $10 + x = 7$ is not solvable in $\mathbb{N}$, $5x = 3$ is not solvable in $\mathbb{Z}$, $x^2 - 2 = 0$ is not solvable in $\mathbb{Q}$, and $x^2 + 1 = 0$ is not solvable in $\mathbb{R}$.

Give an example of an equation that is not solvable in $\mathbb{R}$, but is solvable in $\overline{\mathbb{R}}$.

10.3 Let $a, b, c, d \in \mathbb{R}$. What is the result of multiplying the quaternions $a + bi + cj + dk$ and $a - bi - cj - dk$?

Use your answer to find the multiplicative inverse of $1 + i + j + k$.

10.4 Let M_i, M_j, and M_k be the following matrices:

$$M_i = \begin{bmatrix} 0 & -1 & 0 & 0 \\ 1 & 0 & 0 & 0 \\ 0 & 0 & 0 & -1 \\ 0 & 0 & 1 & 0 \end{bmatrix}, \quad M_j = \begin{bmatrix} 0 & 0 & -1 & 0 \\ 0 & 0 & 0 & 1 \\ 1 & 0 & 0 & 0 \\ 0 & -1 & 0 & 0 \end{bmatrix}, \quad \text{and} \quad M_k = \begin{bmatrix} 0 & 0 & 0 & -1 \\ 0 & 0 & -1 & 0 \\ 0 & 1 & 0 & 0 \\ 1 & 0 & 0 & 0 \end{bmatrix}.$$

Show that these three matrices satisfy the same multiplication rules as the quaternions i, j, and k.

This implies that the quaternion $q = a + bi + cj + dk$ can be represented as the matrix

$$M_q = \begin{bmatrix} a & -b & -c & -d \\ b & a & -d & c \\ c & d & a & -b \\ d & -c & b & a \end{bmatrix} = aI + bM_i + cM_j + dM_k.$$

It follows that for quaternions q and r we have $M_{q+r} = M_q + M_r$ and $M_{q \cdot r} = M_q \cdot M_r$.

Evaluate det M_q.

10.5 Repeat Exercise 10.4 with these 2×2 matrices:

$$M_i = \begin{bmatrix} i & 0 \\ 0 & -i \end{bmatrix}, \quad M_j = \begin{bmatrix} 0 & 1 \\ -1 & 0 \end{bmatrix}, \quad \text{and} \quad M_k = \begin{bmatrix} 0 & i \\ i & 0 \end{bmatrix}.$$

Note that i inside these matrices is the usual $i \in \mathbb{C}$.

10.6 In $\mathbb{Q}_5$ we determined that $-1 = \ldots 444444_{\text{FIVE}}$.

a) What is the 5-adic representation of -2?

b) What is the 5-adic representation of $\frac{1}{2}$?

10.7 This exercise requires some knowledge of how negative integers are stored in a computer using a *two's complement* representation.

a) Write 23 in binary.

b) Write -23 in $\mathbb{Q}_2$.

c) How is -23 represented as a 32-bit integer in a computer?

10.8 Working in $\mathbb{Q}_7$ let $a = \ldots 6421216213_{\text{SEVEN}}$. To the extent possible (since not all digits are given), calculate a^2. What is a?

Hint: Use a computer system that handles very large integers (such as Python). Convert $6421216213_{\text{SEVEN}}$ to base ten, square it, and convert back to base seven.

10.9 In $\mathbb{Q}_5$, let $a = \ldots 223032431212_{\text{FIVE}}$. What is a^2?

10.10 For ordinary decimal numbers, $0.999999\ldots$ is equal to 1. In the ten-adic numbers, $\mathbb{Q}_{10}$, what is the value of $\ldots 999999.0$?

10.11 Working in $\mathbb{Q}_{10}$, the ten-adic numbers, let

$$a = \ldots 9879186432 \quad \text{and} \quad b = \ldots 8212890625.$$

Calculate (as far as reasonable since most of the digits to the left are cut off) $a \cdot b$. Note that neither a nor b is zero. What is $a \cdot b$?

Answers to Exercises

Chapter 0

0.1 An integer n is called a *perfect square* provided there is an integer k such that $n = k^2$.

0.2 The *Fibonacci sequence* is the list of numbers in which the first two terms are 0 and 1, and every term thereafter is the sum of the previous two terms.

Alternatively, the *Fibonacci sequence* is the list $a_0, a_1, a_2, a_3, \ldots$ in which $a_0 = 0$, $a_1 = 1$, and $a_n = a_{n-1} + a_{n-2}$ for all $n \geq 2$.

0.3 A triangle is called *equilateral* if the lengths of its three sides are all equal to each other.

0.4 Three points A, B, and C in the plane are called *collinear* provided there is a line that contains all three.

0.5 Two lines in a plane are *parallel* provided they do not intersect.

Two lines in three-dimensional space are *parallel* provided there is a plane that contains them both and they do not intersect.

0.6 Let a and b be even integers. This means that $a = 2x$ and $b = 2y$ where x and y are integers.

Therefore $ab = (2x)(2y) = 4xy = 2(2xy)$ which shows that ab is 2 times the integer $2xy$ and therefore is even.

0.7 Let a and b be odd integers. This means that $a = 2x + 1$ and $b = 2y + 1$ where x and y are integers.

Their sum $a + b$ is $(2x + 1) + (2y + 1) = 2x + 2y + 2 = 2(x + y + 1)$. Note that $a + b$ is 2 times an integer and is therefore even.

Their product $a \cdot b$ is $(2x + 1) \cdot (2y + 1) = 4xy + 2x + 2y + 1 = 2(2xy + x + y) + 1$. Note that this equals $2k + 1$ where $k = 2xy + x + y$, which is an integer. Therefore ab is odd.

0.8 Let $a = b = \begin{bmatrix} 0 & 1 \\ 0 & 0 \end{bmatrix}$. This is a nonzero matrix, and yet $ab = \begin{bmatrix} 0 & 0 \\ 0 & 0 \end{bmatrix}$ which is the zero matrix.

Chapter 1

1.1 a) Yes. b) No. c) Yes. d) No.

1.2 They are: $\{1,2,3\}, \{1,2\}, \{1,3\}, \{2,3\}, \{1\}, \{2\}, \{3\}$, and $\varnothing$.

1.3 Let $A = \{1, 10\}$ and $B = \{1, 3, 5\}$. Note that $10 \in A$ but $10 \notin B$, and so $A \nsubseteq B$.

1.4 When a set X is not a subset of A it must be the case that X has an element that's not a member of A. Since $\varnothing$ has no elements, that never happens.

Said differently, the requirement is that all of $\varnothing$'s elements must satisfy a condition (being also in A). But since $\varnothing$ has no elements, "all of them" meet the criterion.

1.5 Yes: $\varnothing \subseteq \varnothing$.

1.6 If $A \subseteq B$ and $B \subseteq A$ then it must be the case that $A = B$.

1.7 $A \cap B = A$ exactly when $A \subseteq B$.

1.8 Let $A = \{1, 2, 3\}$, $B = \{3, 4, 5\}$, and $C = \{1, 5, 9\}$. Note that $A \cap B \cap C = \varnothing$ but

$$A \cap B = \{3\}, \quad A \cap C = \{1\}, \quad \text{and} \quad B \cap C = \{5\},$$

and therefore no two of these are disjoint.

1.9 The trick here is that a set may be an element of another set. With this in mind, the simplest answer to this exercise is $A = \{2, \{2\}\}$. This set has two elements: the number 2 and the singleton set $\{2\}$.

1.10 There can be many possible definitions, but here is a relatively simple one:

A set X is a *doubleton* provided $X = A \cup B$ where A and B are singleton sets and $A \neq B$.

1.11 There are 100 ordered pairs of the form (x, y) where $x, y \in A = \{1, 2, \ldots, 10\}$. To see why, note that there are 10 possibilities of the form $(1, y)$, 10 possibilities of the form $(2, y)$, and so forth. Adding these up, we find the number of such ordered pairs is

$$\underbrace{10 + 10 + \cdots + 10}_{10 \text{ terms}} = 10 \cdot 10 = 100.$$

1.12 a) $B - A = \{6, 7\}$.

b) $A - B = B - A$ exactly when $A = B$, in which case both $A - B$ and $B - A$ equal $\varnothing$.

1.13 a) $\{1, 2, 3, 4\} \triangle \{3, 4, 5, 6\} = \{3, 4\}$.

b) $A \triangle A = \varnothing$.

c) $A \triangle \varnothing = A$.

d) If $A \subseteq B$ then $A \triangle B = B - A$.

1.14 a) No, $\oplus$ is not commutative. For example

$$(1, 2) \oplus (3, 4) = (1, 2, 3, 4) \neq (3, 4, 1, 2) = (3, 4) \oplus (1, 2).$$

b) Yes, $\oplus$ is associative.

c) Yes, $\oplus$ has an identity element: the empty list ().

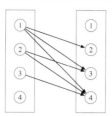

A drawing of the < relation on the set $\{1, 2, 3, 4\}$.

1.15 a) $\{100, 101, 102, \ldots\}$.

b) $\{x \in \mathbb{N} : x \text{ is even}\}$.

c) $\emptyset$.

d) $\{x \in \mathbb{N} : x \neq 4\}$.

1.16 a) When $S = R^{-1}$ then the picture of S is the same as the picture for R except all the arrows are reversed. To clean the picture up further, we would place the set B on the left and the set A on the right.

b) $\left(R^{-1}\right)^{-1} = R$.

c) If R is a one-to-one correspondence from A to B, then $S = R^{-1}$ is a one-to-one correspondence from B to A.

1.17 a) $\{(1, 4), (2, 4), (3, 4), (1, 5), (2, 5), (3, 5)\}$.

b) No, $\times$ is not commutative. For example:

$$\{1, 2\} \times \{3, 4\} = \{(1, 3), (1, 4), (2, 3), (2, 4)\} \quad \text{but}$$
$$\{3, 4\} \times \{1, 2\} = \{(3, 1), (3, 2), (4, 1), (4, 2)\}.$$

Thinking of these two sets as relations, notice that $(A \times B)^{-1} = B \times A$.

c) For the set $\{1, 2, 3, 4\}$, less-than is the set

$$\{(1, 2), (1, 3), (1, 4), (2, 3), (2, 4), (3, 4)\}.$$

d) See the figure on the current page.

1.18 There are six one-to-one correspondences from $A = \{1, 2, 3\}$ to $B = \{4, 5, 6\}$:

$1 \to 4, 2 \to 5, 3 \to 6,$ $1 \to 4, 2 \to 6, 3 \to 5,$
$1 \to 5, 2 \to 4, 3 \to 6,$ $1 \to 5, 2 \to 6, 3 \to 4,$
$1 \to 6, 2 \to 4, 3 \to 5,$ $1 \to 6, 2 \to 5, 3 \to 4.$

1.19 a) The relation is-married-to is symmetric, and the relation is-the-parent-of is not.

b) The relation = (equality) is symmetric and the relation < (less-than) is not.

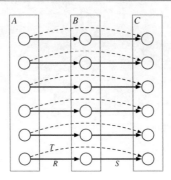

Combining two one-to-one correspondences to give a new one-to-one correspondence.

    c) A relation R is symmetric exactly when $R = R^{-1}$.

1.20 Yes. The relation = is a one-to-one correspondence from A to itself.

1.21 Yes: R^{-1} (see Exercise 1.16 on page 18) simply reverses the arrows of R, giving a one-to-one correspondence from B to A.

1.22 Yes: There is a one-to-one correspondence from A to C. For an element $a \in A$, let b the unique element in B such that $a\ R\ b$. Next let c be the unique element in C such that $b\ S\ c$. Define a new relation T from A to C so that $a\ T\ c$ is the only relation involving a and c. This is illustrated in the figure on this page. It's easy to see that T is a one-to-one correspondence from A to C.

See also Figure 2.1.

1.23   a) Equivalence relation.

    b) Equivalence relation.

    c) Reflexive and transitive, but not symmetric (e.g., $3 \leq 4$ is true but $4 \leq 3$ is false). Not an equivalence relation.

    d) Symmetric, but neither reflexive (you are not your own sibling) nor transitive. For example, Alice and Bob are the parents of child #1, Bob and Carla are the parents of child #2, and Carla and Dave are the parents of child #3. Children #1 and #2 are siblings and children #2 and #3 are siblings, but children #1 and #3 are not.

    e) Equivalence relation.

1.24 There is one partition in which all the parts are singletons:
$$\{\{1\}, \{2\}, \{3\}, \{4\}\}.$$

There are six partitions with one doubleton and two singletons:
$$\{\{1,2\}, \{3\}, \{4\}\},\ \{\{1,3\}, \{2\}, \{4\}\},\ \{\{1,4\}, \{2\}, \{3\}\},$$
$$\{\{2,3\}, \{1\}, \{4\}\},\ \{\{2,4\}, \{1\}, \{3\}\},\ \{\{3,4\}, \{1\}, \{2\}\}.$$

There are three partitions with two doubletons:

$$\{\{1,2\},\{3,4\}\}, \quad \{\{1,3\},\{2,4\}\}, \quad \{\{1,4\},\{2,3\}\}.$$

There are four partitions with one singleton and one triple:

$$\{\{1\},\{2,3,4\}\}, \quad \{\{2\},\{1,3,4\}\}, \quad \{\{3\},\{1,2,4\}\}, \quad \{\{4\},\{1,2,3\}\}.$$

Finally, there is one partition with a single part of size 4:

$$\{\{1,2,3,4\}\}.$$

That accounts for all $1 + 6 + 3 + 4 + 1 = 15$ partitions of $\{1, 2, 3, 4\}$.

1.25   a) Note that (a, b) R (a, b) because $a^2 + b^2 = a^2 + b^2$, and so R is reflexive.
Note that if (a, b) R (c, d) then $a^2 + b^2 = c^2 + d^2$, which we can rewrite as $c^2 + d^2 = a^2 + b^2$ and so (c, d) R (a, b). Therefore R is symmetric.
Finally, if (a, b) R (c, d) and (c, d) R (e, f), then we know that $a^2 + b^2 = c^2 + d^2$ as well as $c^2 + d^2 = e^2 + f^2$. Hence $a^2 + b^2 = e^2 + f^2$, and so (a, b) R (e, f). Thus R is transitive.
Therefore R is an equivalence relation.

     b) The equivalence class $[\![(a, b)]\!]_R$ consists of all points (x, y) for which $x^2 + y^2 = a^2 + b^2$. In other words, $[\![(a, b)]\!]_R$ is a circle of radius $\sqrt{a^2 + b^2}$.
Note that $[\![(0, 0)]\!]_R$ is a "degenerate" circle whose radius is zero; that is, $[\![(0, 0)]\!]_R = \{(0, 0)\}$.

1.26 $[\![t]\!]_\sim$ is the set of all isosceles right triangles (with angles 90°, 45°, and 45°).

1.27   a) R is reflexive: For any integer, x, we have x R x because $0 = x - x$ is even.
R is symmetric: For any two integers, x and y, if x R y we know that $x - y$ is even. This implies $-(x - y) = y - x$ is even and so y R x.
R is transitive: For any three integers, x, y, and z, suppose x R y and y R z. This means that both $x - y$ and $y - z$ are even. Adding these, we have $(x - y) + (y - z) = x - z$ is even (because the sum of even numbers is even) and so x R z.
Therefore R is an equivalence relation.

     b) There are two equivalence classes: $[\![0]\!]_R$ contains all the even integers and $[\![1]\!]_R$ contains all the odd integers.

1.28   a) R is not reflexive: For example, 2 R 2 is false because $2 - 2$ is not odd.
R is symmetric: For integers x and y, if x R y then $x - y$ is odd. This implies that $-(x - y) = y - x$ is odd and therefore y R x.
R is not transitive: For example 1 R 2 (because $2 - 1$ is odd) and 2 R 3 (because $3 - 2$ is odd), but it is not true that 1 R 3 (because $3 - 1$ is not odd).
Therefore R is not an equivalence relation.

     b) Since R is not an equivalence relation, there are no equivalence classes.

1.29 a) The first condition in the definition implies that there is an arrow emerging from every element of a, and the second condition implies there is at most one arrow emerging from a. Thus, for every a in A there is exactly one arrow from that element pointing to some member of B.

b) If f is a function, then f^{-1} is also a function exactly when f is a one-to-one correspondence from A to B.

1.30 a) $5 \star 3 = 5^2 - 3^2 = 25 - 9 = 16$.

b) No, $-$ is not an operation on $\mathbb{N}$ because, by our definition, for any two natural numbers a and b, it must be the case that $a - b$ is a natural number. However, $3 - 5$ is not a natural number.

c) An operation $\star$ on the set $\{1,2\}$ looks like this:

$$1 \star 1 = a, \quad 1 \star 2 = b, \quad 2 \star 1 = c, \quad 2 \star 2 = d$$

where a, b, c, d can be any values in $\{1, 2\}$. Writing out all the possibilities of (a, b, c, d) from $(1, 1, 1, 1)$ to $(2, 2, 2, 2)$ we find there are 16 possible operations.

1.31 An operation cannot have two different identity elements; here's why. Let's say that e_1 and e_2 are identity elements for $\star$. What is $e_1 \star e_2$?

On the one hand, since e_1 is an identity element, $e_1 \star e_2 = e_2$. On the other hand, since e_2 is an identity element, $e_1 \star e_2 = e_1$. Writing these on a single line we have

$$e_1 = e_1 \star e_2 = e_2$$

and therefore $e_1 = e_2$. So the two "different" identity elements must, in fact, be equal.

Chapter 2

2.1 The set $\{a, b, c\}$ and $\{b, c, d\}$ (where a, b, c, and d are different basic objects) are representatives of **3**, but they are not disjoint because $\{a, b, c\} \cap \{b, c, d\} = \{b, c\} \neq \varnothing$.

The sets $\{a, b, c\}$ and $\{x, y, z\}$ (where a, b, c, x, y, and z are different basic objects) are disjoint representatives of **3**.

2.2 This is true. The $\equiv$ equivalence class of the empty set, $\varnothing$, consists of all sets that have a one-to-one correspondence with $\varnothing$. The only set that has such a correspondence with the empty set is the empty set itself, $\varnothing$. Hence $[\![\varnothing]\!] = \{\varnothing\}$.

2.3 If A and B are infinite, then $A \cup B$ must be infinite, but it might be the case that $A \cap B$ is finite. For example, consider these infinite sets: let A be the set of even natural numbers and let B be the set of primes. Note that $A \cap B = \{2\}$ which is finite.

2.4 This is true. Remember that **a** is an equivalence class of sets that have one-to-one correspondences with each other. That is, $\mathbf{a} = [\![A]\!]$.

2.5 a) $\mathbf{a} + \mathbf{b}$.   b) **0**.   c) **a**.   d) $\mathbf{a} \cdot \mathbf{b}$.

2.6 Note that

$$\{a,b\} \times \{c,d,e\} = \{(a,c),(a,d),(a,e),(b,c),(b,d),(b,e)\} \quad \text{and}$$
$$\{c,d,e\} \times \{a,b\} = \{(c,a),(c,b),(d,a),(d,b),(e,a),(e,b)\}.$$

Here is a one-to-one correspondence between these sets:

$$(a,c) \to (c,a), \qquad (a,d) \to (d,a), \qquad (a,e) \to (e,a),$$
$$(b,c) \to (c,b), \qquad (b,d) \to (d,b), \qquad (b,e) \to (e,b).$$

2.7 a) Let a,b,c,d,a',b',c',d' be distinct objects. Observe that $\{a,b,c,d\}$ and $\{a',b',c',d'\}$ are disjoint representatives of **4**. Therefore their union is a representative of $\mathbf{4+4}$:

$$S = \{a,b,c,d,a',b',c',d'\} \in \mathbf{4+4}.$$

b) Now let x and y be distinct objects, so $\{x,y\}$ is a representative of **2**. Then

$$T = \{x,y\} \times \{a,b,c,d\}$$
$$= \{(x,a),(x,b),(x,c),(x,d),(y,a),(y,b),(y,c),(y,d)\}$$

is a representative of $\mathbf{2 \cdot 4}$.

c) We have the following one-to-one correspondence between these two sets:

$$(x,a) \to a, \qquad (x,b) \to b, \qquad (x,c) \to c, \qquad (x,d) \to d,$$
$$(y,a) \to a', \qquad (y,b) \to b', \qquad (y,c) \to c', \qquad (y,d) \to d'.$$

d) Therefore $\{a,b,c,d\} \cup \{a',b',c',d'\} \equiv \{x,y\} \times \{a,b,c,d\}$ and so $\mathbf{4+4} = \mathbf{2 \cdot 4}$.

2.8 a) Let A and A' be disjoint representatives of **a**. Then $A \cup A'$ is a representative of $\mathbf{a+a}$. Let $S = A \cup A'$.

b) Let $\{x,y\}$ be a representative of **2** and A be a representative of **a** (same A as above). Then $T = \{x,y\} \times A$ is a representative of $\mathbf{2 \cdot a}$.

c) Since A and A' are both members of the equivalence class **a** there is a one-to-one correspondence between them. For $a \in A$, let a' be the element of A' to which a corresponds; that is, $a \to a'$.

Elements of $T = \{x,y\} \times A$ are either of the form (x,a) or (y,a) where $a \in A$. Here is a one-to-one correspondence from S to T:

$$a \to (x,a) \quad \text{and} \quad a' \to (y,a).$$

d) We have shown that $S \equiv T$ where S is a representative of $\mathbf{a+a}$ and T is a representative of $\mathbf{2 \cdot a}$. Therefore $\mathbf{a+a} = [\![S]\!] = [\![T]\!] = \mathbf{2 \cdot A}$.

2.9 No, $[\![\{a\}]\!]$ is not a finite set. There are infinitely many possible singleton sets and they are elements of $[\![\{a\}]\!]$.

178   From Counting to Continuum

2.10 Let A be a representative of **a** and let B be a representative of **b** that is disjoint from A. In other words,
$$\mathbf{a} = [\![A]\!] \quad \text{and} \quad \mathbf{b} = [\![B]\!] \quad \text{where} \quad A \cap B = \emptyset.$$
By definition, $\mathbf{a} + \mathbf{b} = [\![A \cup B]\!]$. Since $A \subseteq A \cup B$, it follows from Definition 2.7 that $[\![A]\!] \leq [\![A \cup B]\!]$, and that's exactly $\mathbf{a} \leq \mathbf{a} + \mathbf{b}$.

2.11 Definition 2.7 tells us that **a** has a representative that is in one-to-one correspondence with a subset of B. Call that subset A. Therefore $A \in \mathbf{a}$ and so A is a representative of **a** that is a subsets of B.

2.12 a) We have
$$A \times B = \{(a,c), (a,d), (a,e), (b,c), (b,d), (b,e)\},$$
$$A \times C = \{(a,x), (a,y), (b,x), (b,y)\}.$$

b) $(A \times B) \cup (A \times C)$ is
$$\{(a,c), (a,d), (a,e), (a,x), (a,y), (b,c), (b,d), (b,e), (b,x), (b,y)\}.$$

c) $B \cup C = \{c, d, e, x, y\}$.

d) We have
$$A \times (B \cup C) = \{a, b\} \times \{c, d, e, x, y\}$$
$$= \{(a,c), (a,d), (a,e), (a,x), (a,y),$$
$$(b,c), (b,d), (b,e), (b,x), (b,y)\}.$$

e) The two sets $(A \times B) \cup (A \times C)$ and $A \times (B \cup C)$ are the same.

2.13 By Definition 2.7, if $\mathbf{a} \leq \mathbf{b}$ we can find $A \in \mathbf{a}$ and $B \in \mathbf{b}$ with $A \subseteq B$. Let $C = B - A$. Define $\mathbf{b} - \mathbf{a}$ to be $[\![C]\!]$.

2.14 Yes. By definition, if x and y are basic objects, then so is the pair (x, y). Since $A \times B = \{(a,b) : a \in A, b \in B\}$, the elements of $A \times B$ are pairs of basic objects, which are themselves basic objects.

2.15 We have the familiar algebraic identity $(\mathbf{a} + \mathbf{b})^2 = \mathbf{a}^2 + 2\mathbf{ab} + \mathbf{b}^2$.

2.16 Look at the vertical **a**-by-**b** rectangle (highlighted in Figure 2.9 on page 49) and the analogous one running horizontally at the bottom of the figure. The number of squares in those two rectangles combined is $2\mathbf{ab}$ where we have counted the squares in the **b**-by-**b** square at the lower right twice. This is at most the number of squares in the full **a**-by-**a** checkerboard plus a second copy of the lower right **b**-by-**b** board. This gives $2\mathbf{ab} \leq \mathbf{a}^2 + \mathbf{b}^2$.

2.17 Given < we define as the other relations as follows:
- $\mathbf{a} \leq \mathbf{b}$ means $\mathbf{a} < \mathbf{b}$ or $\mathbf{a} = \mathbf{b}$.
- $\mathbf{a} > \mathbf{b}$ means $\mathbf{b} < \mathbf{a}$.
- $\mathbf{a} \geq \mathbf{b}$ means $\mathbf{b} < \mathbf{a}$ or $\mathbf{b} = \mathbf{a}$.

2.18 If $a = b$ then neither $a < b$ nor $a > b$ can be true because, by definition, the relations $<$ and $>$ are false for equal numbers.

Otherwise $a \ne b$. By (L2) either $a \le b$ or $b \le a$. In the first case, we have $a < b$ and in the second case $a > b$. But can we have both? That would imply that $a \le b$ and $b \le a$, and by (L3) we have $a = b$, which is false.

2.19 Every $a \in \mathbb{N}$ divides 0 because $a \cdot 0 = 0$. On the other hand, the only number a so that $0|a$ is $a = 0$.

2.20 **Reflexive**: To show that $|$ is reflexive we need to show that $a|a$ for all $a \in \mathbb{N}$. This is so because $a = 1 \cdot a$.

Antisymmetric: To show that $|$ is antisymmetric we need to show that if $a|b$ and $b|a$, then $a = b$. This means there are natural numbers x and y so that $a = bx$ and $b = ay$. If one of a or b is 0, the equations $a = bx$ and $b = ay$ imply that both must be zero, and so $a = b$.

If neither a nor b is 0, then $a = xb = x(ya) = (xy)a$. This implies that $xy = 1$ and, since x and y are natural numbers, we must have $x = y = 1$. Substituting $x = 1$ in $a = xb$ we conclude that $a = b$.

Transitive: We are given that $a|b$ and $b|c$. That $a|b$ means there is an $x \in \mathbb{N}$ such that $b = ax$. That $b|c$ means there is a $y \in \mathbb{N}$ such that $c = by$.

Thus $c = by = (ax)y = a(xy)$. Therefore $a|c$.

2.21 a) **2**. b) **4**. c) **1**. d) **10**.

2.22 Suppose there were such an infinite descending list of natural numbers, $a_1 > a_2 > a_3 > \cdots$.

Let A be the set of numbers on the list. This is a nonempty subset of $\mathbb{N}$. Therefore A has a least element, b. Since b is on the list, it's at some position, say, position n. Thus $b = a_n$. But then a_{n+1} is also in A and $a_{n+1} < b$, a contradiction. Therefore no such infinite descending list exists.

2.23 The sums are as follows: a) **12**, b) **17**, c) **2**, d) **0**.

The products are as follows: a) **48**, b) **210**, c) **2**, d) **1**.

2.24 Let n be a natural number. Then, by Proposition 2.8, we can write $n = 2a + r$ where a is a natural number and $0 \le r < 2$. In other words, $r = 0$ or $r = 1$.

We are given that n is not even, so $r \ne 0$. Therefore $n = 2a + 1$.

2.25 $1 = \text{next}(0)$.

2.26 For $n \in \mathbb{N}$ define:

- $n \cdot 0$ is 0, and
- $n \cdot \text{next}(a)$ is $n \cdot a + n$.

For example, $n \cdot 2 = n \cdot \text{next}(1) = n \cdot 1 + n$ and $n \cdot 1 = n \cdot \text{next}(0) = n \cdot 0 + n = 0 + n = n$. Combining these, $n \cdot 2 = n + n$. Whew!

2.27 a) $123_{\text{FIVE}} = 38_{\text{TEN}}$ and $123_{\text{SEVEN}} = 66_{\text{TEN}}$.

b) $145_{\text{TEN}} = 1040_{\text{FIVE}} = 265_{\text{SEVEN}}$.

c) 125_{SEVEN}.

d) 102_{FIVE}.

2.28 The answer is no because we defined **1** to be the equivalence class of all singleton sets of *basic objects* and **1** is not a basic object; it's something way more complicated.

We might be tempted to define natural numbers to be equivalence classes of all finite sets irrespective of the basicness of their elements. Then we would not need to worry about which things are basic and which are not.

Alas, this approach would pull us into Bertrand Russell's paradox that there is no set of all sets.

Logicians – mathematicians who work on the foundations of mathematics – provide a way out of this pickle.

2.29 One primordial object is enough! We could have defined a basic object like this: An object Z is a *basic object* if either $Z = a$ or $Z = (x, y)$ where x and y are basic objects.

Then we'd start generating basic stuff starting with a and (a, a), and then we'd have
$$(a, (a, a)), \quad ((a, a), a), \quad \text{and} \quad ((a, a), (a, a)),$$
and so forth.

Starting with two primordial basic objects is a little less scary.

2.30 Yes: If A and B are sets of basic objects, then the same is true for $A \times B$ because ordered pairs of basic objects are also basic objects.

The point of this exercise is to ensure that if **a** and **b** are natural numbers, then there are sets of basic objects in the class $\mathbf{a} \cdot \mathbf{b}$.

Chapter 3

3.1 $[\![(3,9)]\!] = [\![(5,11)]\!]$ because $3 + 11 = 14 = 9 + 5 \Rightarrow (3,9) \equiv (5,11)$.

3.2
$$5 = [\![(5,0)]\!] = \{(5,0), (6,1), (7,2), (8,3), \ldots\},$$
$$-5 = [\![(0,5)]\!] = \{(0,5), (1,6), (2,7), (3,8), \ldots\}.$$

3.3 Let $a = (6, 1)$ and $b = (9, 4)$ be representatives of **5**, and let $(2, 5)$ and $(6, 9)$ be representatives of **−3**.

For addition, we have the following:
$$a + c = (6,1) + (2,5) = (8,6) \equiv (2,0),$$
$$a + d = (6,1) + (6,9) = (12,10) \equiv (2,0),$$
$$b + c = (9,4) + (2,5) = (11,9) \equiv (2,0),$$
$$b + d = (9,4) + (6,9) = (15,13) \equiv (2,0),$$

illustrating that $5 + (-3) = 2$.

For multiplication, we have the following:

$$a \otimes c = (6,1) \otimes (2,5) = (17, 32) \equiv (0, 15),$$
$$a \otimes d = (6,1) \otimes (6,9) = (45, 60) \equiv (0, 15),$$
$$b \otimes c = (9,4) \otimes (2,5) = (38, 53) \equiv (0, 15),$$
$$b \otimes d = (9,4) \otimes (6,9) = (90, 105) \equiv (0, 15),$$

illustrating that $(5) \cdot (-3) = -15$.

3.4 Integers $[\![(a,b)]\!]$ and $[\![(b,a)]\!]$ are negatives of each other:

$$[\![(a,b)]\!] + [\![(b,a)]\!] = [\![(a+b, a+b)]\!] = [\![(0,0)]\!] = \mathbf{0}.$$

3.5 We have $(a,b) \otimes (a,b) = (a^2+b^2, 2ab)$ and $(a,b) \otimes (b,a) = (2ab, a^2+b^2)$. This illustrates that for $\mathbf{x} \in \mathbb{Z}$ that $\mathbf{x}^2$ is the negative of $(\mathbf{x}) \cdot (-\mathbf{x})$.

3.6 + is commutative.

Let $\mathbf{x} = [\![(a,b)]\!]$ and $\mathbf{y} = [\![(c,d)]\!]$. Then

$$\mathbf{x} + \mathbf{y} = [\![(a,b)]\!] + [\![(c,d)]\!]$$
$$= [\![(a+c, b+d)]\!]$$
$$= [\![(c+a, d+b)]\!]$$
$$= [\![(c,d)]\!] + [\![(a,b)]\!] = \mathbf{y} + \mathbf{x}.$$

+ is associative.

Let $\mathbf{x} = [\![(a,b)]\!]$, $\mathbf{y} = [\![(c,d)]\!]$, and $\mathbf{z} = [\![(e,f)]\!]$. Then

$$(\mathbf{x} + \mathbf{y}) + \mathbf{z} = [\![(a+c, b+d)]\!] + [\![(e,f)]\!]$$
$$= [\![((a+c)+e), ((b+d)+f)]\!]$$
$$= [\![(a+(c+e)), (b+(d+f))]\!]$$
$$= [\![(a,b)]\!] + [\![(c+e, d+f)]\!] = \mathbf{x} + (\mathbf{y} + \mathbf{z}).$$

3.7 a) $\mathbf{2} = \{(n+2, n) : n \in \mathbb{N}\}$.

b) Let $(a+1, a)$ and $(b+1, b)$ be representatives of $\mathbf{1}$. By Definition 3.4, $\mathbf{1} + \mathbf{1}$ is the equivalence class of $((a+1) + (b+1), a+b) = ((a+b)+2, (a+b))$. Taking $n = a+b$, we see that this is an element of $\mathbf{2}$ and therefore $\mathbf{1} + \mathbf{1} = \mathbf{2}$.

c) Let $(a+1, a)$ and $(b+1, b)$ be representatives of $\mathbf{1}$. By Definition 3.5,

$$\mathbf{1} \cdot \mathbf{1} = [\![((a+1)(b+1) + ab, (a+1)b + a(b+1))]\!]$$
$$= [\![(ab + a + b + 1 + ab, ab + b + ab + a)]\!]$$
$$= [\![(2ab + a + b + 1, 2ab + a + b)]\!] = \mathbf{1}.$$

3.8 Let (n,n) be a representative of $\mathbf{0}$ and let (a,b) be a representative of $\mathbf{x}$. Then, applying Definition 3.5, we have

$$\mathbf{0} \cdot \mathbf{x} = [\![(na + nb, nb + na)]\!] = \mathbf{0}.$$

3.9 Let (a, b) be a representative of **x** and let $(n + 2, n)$ be a representative of **2**. Then

$$\mathbf{2 \cdot x} = [\![((n+2)a + nb, (n+2)b + na)]\!]$$
$$= [\![(na + nb + 2a, na + nb + 2b)]\!]$$
$$= [\![(2a, 2b)]\!].$$

This result cannot be **1** because it would require $2a$ to be exactly 1 greater than $2b$. Since $2a$ is even, it cannot be one greater than another even natural number.

3.10 $\otimes$ is commutative.

Let $a, b, c, d \in \mathbb{N}$. We have

$$(a, b) \otimes (c, d) = (ac + bd, ad + bc),$$
$$(c, d) \otimes (a, b) = (ca + db, cb + da) = (ac + bd, ad + bc),$$

and therefore $(a, b) \otimes (c, d) = (c, d) \otimes (a, b)$.

$\otimes$ is associative.

Let $a, b, c, d, e, f \in \mathbb{N}$. We have

$$\Big((a, b) \otimes (c, d)\Big) \otimes (e, f) = (ac + bd, ad + bc) \otimes (e, f)$$
$$= \Big((ac + bd)e + (ad + bc)f, (ac + bd)f + (ad + bc)e\Big)$$
$$= (ace + bde + adf + bcf, acf + bdf + ade + bce),$$

$$(a, b) \otimes \Big((c, d) \otimes (e, f)\Big) = (a, b) \otimes (ce + df, cf + de)$$
$$= \Big(a(ce + df) + b(cf + de), a(cf + de) + b(ce + df)\Big)$$
$$= (ace + adf + bcf + bde, acf + ade + bce + bdf)$$
$$= (ace + bde + adf + bcf, acf + bdf + ade + bce),$$

and therefore $\Big((a, b) \otimes (c, d)\Big) \otimes (e, f) = (a, b) \otimes \Big((c, d) \otimes (e, f)\Big)$.

3.11 Let's choose $(6, 1)$ and $(10, 5)$ as representatives of **5** and $(10, 2)$ and $(12, 4)$ as representatives of **8**. We verify that $\mathbf{5 \leq 8}$ four times like this:

- $[\![(6, 1)]\!] \leq [\![(10, 2)]\!]$ because $6 + 2 \leq 1 + 10$,
- $[\![(6, 1)]\!] \leq [\![(12, 4)]\!]$ because $6 + 4 \leq 1 + 12$,
- $[\![(10, 5)]\!] \leq [\![(10, 2)]\!]$ because $10 + 2 \leq 5 + 10$, and
- $[\![(10, 5)]\!] \leq [\![(12, 4)]\!]$ because $10 + 4 \leq 5 + 12$.

3.12 We showed that **0** times any integer is **0**. So if $\mathbf{x \cdot y \neq 0}$, it's impossible that either factor is **0** because otherwise the result would be **0**.

3.13 a) Add $-\mathbf{x}$ to both sides:
$$\mathbf{a} + \mathbf{x} = \mathbf{b} + \mathbf{x}$$
$$\Rightarrow \quad (\mathbf{a} + \mathbf{x}) + (-\mathbf{x}) = (\mathbf{b} + \mathbf{x}) + (-\mathbf{x})$$
$$\Rightarrow \quad \mathbf{a} + (\mathbf{x} + -\mathbf{x}) = \mathbf{b} + (\mathbf{x} + -\mathbf{x})$$
$$\Rightarrow \quad \mathbf{a} + \mathbf{0} = \mathbf{b} + \mathbf{0}$$
$$\Rightarrow \quad \mathbf{a} = \mathbf{b}.$$

b) Let $\mathbf{a} = (a, b)$, $\mathbf{b} = (c, d)$, and $\mathbf{x} = (x, y)$. We are given that $\mathbf{a} \cdot \mathbf{x} = \mathbf{b} \cdot \mathbf{x}$, which we can rewrite as
$$\mathbf{a} \cdot \mathbf{x} = [\![(a, b) \otimes (x, y)]\!] = [\![(ax + by, ay + bx)]\!],$$
$$\mathbf{b} \cdot \mathbf{x} = [\![(c, d) \otimes (x, y)]\!] = [\![(cx + dy, cy + dx)]\!],$$
and therefore $(ax + by, ay + bx) \equiv (cx + dy, cy + dx)$, which gives
$$(ax + by) + (cy + dx) = (ay + bx) + (cx + dy)$$
$$\Rightarrow \quad (a + d)x + (b + c)y = (b + c)x + (a + d)y.$$

Observe that
$$(a + d, b + c) \otimes (x, y) = \big((a + d)x + (b + c)y, (a + d)y + (b + c)x\big) \equiv \mathbf{0}.$$

Since $\mathbf{x} \neq \mathbf{0}$ and $[\![(a + d, b + c)]\!] \cdot \mathbf{x} = \mathbf{0}$, it follows by Exercise 3.12 that $[\![(a + d, b + c)]\!] = \mathbf{0}$. This implies that $a + d = b + c$, whence $(a, b) \equiv (c, d)$, and we conclude $\mathbf{a} = \mathbf{b}$.

3.14 Let $\mathbf{x} = [\![(a, b)]\!]$, $\mathbf{y} = [\![(c, d)]\!]$, and $\mathbf{z} = [\![(e, f)]\!]$. We calculate:
$$\mathbf{x} \cdot (\mathbf{y} + \mathbf{z}) = [\![(a, b)]\!] \cdot \Big([\![(c, d)]\!] + [\![(e, f)]\!]\Big)$$
$$= [\![(a, b)]\!] \cdot [\![(c + e, d + f)]\!]$$
$$= [\![a(c + e) + b(d + f), a(d + f) + b(c + e)]\!]$$
$$= [\![(ac + ae + bd + bf, ad + af + bc + be)]\!]$$

and
$$\mathbf{x} \cdot \mathbf{y} + \mathbf{x} \cdot \mathbf{z} = [\![(a, b)]\!] \cdot [\![(c, d)]\!] + [\![(a, b)]\!] \cdot [\![(e, f)]\!]$$
$$= [\![(ac + bd, ad + bc)]\!] + [\![(ae + bf, af + be)]\!]$$
$$= \Big[\!\!\Big[\big((ac + bd) + (ae + bf), (ad + bc) + (af + be)\big)\Big]\!\!\Big]$$
$$= [\![(ac + ae + bd + bf, ad + af + bc + be)]\!] = \mathbf{x} \cdot (\mathbf{y} + \mathbf{z}).$$

3.15 Let $\mathbf{x}$ and $\mathbf{y}$ be positive integers. This means they can be represented as $\mathbf{x} = [\![(a, 0)]\!]$ and $\mathbf{y} = [\![(b, 0)]\!]$ where a and b are nonzero natural numbers. Then
$$\mathbf{x} \cdot \mathbf{y} = [\![(a \cdot b + 0 \cdot 0, a \cdot 0 + 0 \cdot b)]\!] = [\![(ab, 0)]\!],$$
which is positive.

If $\mathbf{x}$ and $\mathbf{y}$ are negative we represent them as $\mathbf{x} = [\![(0, a)]\!]$ and $\mathbf{y} = [\![(0, b)]\!]$ where, as before, a and b are nonzero natural numbers. Then
$$\mathbf{x} \cdot \mathbf{y} = [\![(0 \cdot 0 + a \cdot b, 0 \cdot b + a \cdot 0)]\!] = [\![(ab, 0)]\!],$$

which is positive.

However, if $\mathbf{x} = [\![(a,0)]\!]$ is positive and $\mathbf{y} = [\![(0,b)]\!]$ is negative, then

$$\mathbf{x} \cdot \mathbf{y} = [\![(a \cdot 0 + 0 \cdot b, a \cdot b + 0 \cdot 0)]\!] = [\![(0, ab)]\!],$$

which is negative.

3.16 If $\mathbf{x}$ and $\mathbf{y}$ are positive, then $\mathbf{x} = [\![(a,0)]\!]$ and $\mathbf{y} = [\![(b,0)]\!]$ for nonzero natural numbers a and b. Then $\mathbf{x} + \mathbf{y} = [\![(a+b,0)]\!]$, which is also positive.

If $\mathbf{x}$ and $\mathbf{y}$ are negative, then we have $\mathbf{x} = [\![(0,a)]\!]$ and $\mathbf{y} = [\![(0,b)]\!]$. Then $\mathbf{x} + \mathbf{y} = [\![(0, a+b)]\!]$, which is also negative.

3.17 The addition properties (A1)–(A5), multiplication properties (M1)–(M4), and the distributive property all apply to $\mathbb{Z}$ as well as $\mathbb{N}$.

The four ordering properties (L1)–(L4) also apply to $\mathbb{Z}$, as does the property (LA).

However, (LM) does not apply to the integers. For example, take $\mathbf{a} = 3$, $\mathbf{b} = 5$, and $\mathbf{c} = -1$. Note that $\mathbf{a} \leq \mathbf{b}$ (i.e., $3 \leq 5$) but it is not the case that $\mathbf{a} \cdot \mathbf{c} \leq \mathbf{b} \cdot \mathbf{c}$ (i.e., it is not true that $-3 \leq -5$).

Finally, (WO) does not apply to the integers. For example, let A be the set of even integers. It is a nonempty subset of $\mathbb{Z}$, but it does not have a least element.

3.18 Yes, $\mathbb{Z}[x]$ is a commutative ring, but there is no reasonable way to define $\leq$ for $\mathbb{Z}[x]$ that interacts appropriately with $+$ and $\cdot$.

Chapter 4

4.1 In $\mathbb{Z}_9$ the additive inverse of **2** is **7** and the additive inverse of **6** is **3**.

The multiplicative inverse of **2** is **5**. However, **6** does not have a multiplicative inverse.

4.2 Let $m = 12$ and note that $\mathbf{3} \cdot \mathbf{4} = \mathbf{0}$ in $\mathbb{Z}_{12}$.

In general, if m is composite with $m = ab$ where $1 < a, b < m$, then $\mathbf{a} \cdot \mathbf{b} = \mathbf{0}$ in $\mathbb{Z}_m$.

4.3 For the relation $\equiv$, congruence modulo m, the following hold:

- $\equiv$ is reflexive: Let $a \in \mathbb{Z}$, then $a \equiv a$ because $a - a = 0$ and $m | 0$ (clearly $0 = m \cdot 0$).

- $\equiv$ is symmetric: Suppose $a \equiv b$. This means that $m | (a - b)$, which means that $a - b = tm$ for some integer t. It follows that $b - a = (-t)m$ and thus $b - a$ is divisible by m, and so $b \equiv a$.

- $\equiv$ is transitive: Suppose $a \equiv b$ and $b \equiv c$. This gives:

$$a \equiv b \implies m|(a-b) \implies a - b = sm,$$
$$b \equiv c \implies m|(b-c) \implies b - c = tm$$

for some integers s and t. Adding the two equations on the right gives

$$a - c = (a - b) + (b - c) = sm + tm = (s+t)m.$$

Thus $m|(a-c)$ and we conclude $a \equiv c$.

4.4 With $m = 2$, $[\![0]\!]$ are the *even* numbers and $[\![1]\!]$ are the *odd* numbers.

4.6 When $m = 1$, all integers are congruent to each other because all integers are divisible by 1. This means there is only one equivalence class, $[\![0]\!]$, that contains all the integers. The operations of addition and multiplication work fine. The only issue is that this does not yield a commutative ring because in $\mathbb{Z}_1$ we have $\mathbf{0} = \mathbf{1}$.

4.7 The best way to do this is to write a program that checks if squaring a number between 1 and $p-1$ gives $p-1$ when reduced mod p. For odd primes less than 100, the result looks like this:

3	none
5	2
7	none
11	none
13	5
17	4
19	none
23	none
29	12
31	none
37	6
41	9
43	none
47	none
53	23
59	none
61	11
67	none
71	none
73	27
79	none
83	none
89	34
97	22

Notice that if $p \equiv 1 \pmod 4$ then -1 is a quadratic residue modulo p, but when $p \equiv 3 \pmod 4$ it is not.

Chapter 5

5.1 a) True. b) True. c) False.

5.2 $[\![(a,b)]\!] + [\![(a,b)]\!] = [\![(ab+ba, b^2)]\!] = [\![(2ab, b^2)]\!] = [\![(2a, b)]\!]$.
$[\![(a,b)]\!] \cdot [\![(a,b)]\!] = [\![(a^2, b^2)]\!]$.

5.3 Let (a,b) be a representative of $\mathbf{x}$ and (c,d) be a representative of $\mathbf{y}$.
Define $\mathbf{x} - \mathbf{y}$ to be $[\![(ad-bc, bd)]\!]$ and $\mathbf{x} \div \mathbf{y} = [\![(ad, bc)]\!]$.

5.4 We are given $(a,b) \equiv (a',b')$ and $(c,d) \equiv (c',d')$. This means that $ab' = a'b$ and $cd' = c'd$.
To show that $(ad+bc, bd) \equiv (a'd'+b'c', b'd')$ we must have
$$(ab+bc)b'd' = (a'd'+b'c')bd,$$

which is equivalent to
$$abb'd' + bcb'd' = a'd'bd + b'c'bd. \qquad (*)$$
We need to verify $(*)$. To do that we rewrite the two terms on the left-hand side
$$abb'd' = (ab')(bd') = (a'b)(bd') = a'd'bd,$$
$$bcb'd' = (b'c)(bd') = (bc')(b'd) = b'c'bd$$
and note they match the two terms on the right. Hence $(*)$ is true and we have verified $(ad+bc, bd) \equiv (a'd' + b'c', b'd')$.

Next we check that $(ac, bd) \equiv (a'c', b'd')$. To do this, we must check that $acb'd' = a'c'bd$. This follows because
$$acb'd' = (ab')(cd') = (a'b)(c'd) = a'c'bd.$$

5.5 Let $\mathbf{x} = [\![(a,b)]\!]$, $\mathbf{y} = [\![(c,d)]\!]$, and $\mathbf{z} = [\![(e,f)]\!]$.

For addition:
$$\begin{aligned}(\mathbf{x} + \mathbf{y}) + \mathbf{z} &= [\![(ad + bc, bd)]\!] + [\![(e, f)]\!] \\ &= [\![((ad+bc)f + bde, (bd)f)]\!] \\ &= [\![(adf + bcf, bdf)]\!]\end{aligned}$$

and
$$\begin{aligned}\mathbf{x} + (\mathbf{y} + \mathbf{z}) &= [\![(a, b)]\!] + [\![(cf + de, df)]\!] \\ &= [\![\big(a(df) + b(cf + de), b(df)\big)]\!] \\ &= [\![(adf + bcf + bde, bdf)]\!],\end{aligned}$$

and therefore $(\mathbf{x} + \mathbf{y}) + \mathbf{z} = \mathbf{x} + (\mathbf{y} + \mathbf{z})$.

For multiplication:
$$\begin{aligned}(\mathbf{x} \cdot \mathbf{y}) \cdot \mathbf{z} &= [\![(ac, bd)]\!] \cdot [\![(e, f)]\!] \\ &= [\![(acf, bde)]\!] \\ &= [\![(a, b)]\!] \cdot [\![(cf, de)]\!] \\ &= \mathbf{x} \cdot (\mathbf{y} \cdot \mathbf{z}).\end{aligned}$$

5.6 Let $\mathbf{x} = [\![(a,b)]\!]$, $\mathbf{y} = [\![(c,d)]\!]$, and $\mathbf{z} = [\![(e,f)]\!]$. We calculate:
$$\begin{aligned}\mathbf{x} \cdot (\mathbf{y} + \mathbf{z}) &= [\![(a, b)]\!] \cdot [\![(cf + de, df)]\!] \\ &= [\![(acf + ade, bdf)]\!]\end{aligned}$$

$$\begin{aligned}\mathbf{x} \cdot \mathbf{y} + \mathbf{x} \cdot \mathbf{z} &= [\![(ac, bd)]\!] + [\![(ae, bf)]\!] \\ &= [\![(ac \cdot bf + bd \cdot ae, bd \cdot bf)]\!] \\ &= [\![\big(b(acf + ade), b(bdf)\big)]\!].\end{aligned}$$

To show that $\mathbf{x} \cdot (\mathbf{y} + \mathbf{z}) = \mathbf{x} \cdot \mathbf{y} + \mathbf{x} \cdot \mathbf{z}$ we must verify that
$$(acf + ade, bdf) \equiv \big(b(acf + ade), b(bdf)\big).$$

By calculating

$$(acf + ade) \cdot b(bdf) \quad \text{and} \quad (bdf) \cdot b(acf + ade),$$

we see that both of these expressions equal

$$ab^2cdf^2 + ab^2d^2ef.$$

5.7 $\mathbb{Z}_6$ is not a field because 2 does not have a multiplicative inverse:

$$2 \cdot 0 = 0, \quad 2 \cdot 1 = 2, \quad 2 \cdot 2 = 4, \quad 2 \cdot 3 = 0, \quad 2 \cdot 4 = 2, \quad 2 \cdot 5 = 4.$$

However, $\mathbb{Z}_7$ is a field. Here are the inverses of the nonzero elements:

$$1^{-1} = 1, \quad 2^{-1} = 4, \quad 3^{-1} = 5, \quad 4^{-1} = 2, \quad 5^{-1} = 3, \quad 6^{-1} = 6.$$

$(\mathbb{Z}_m, +, \cdot)$ is a field exactly when m is prime. See, for example, [10].

5.8 We know that $(\mathbb{Z}_5, +, \cdot)$ is a field.

The suggested order relation satisfies the reflexive, antisymmetric, transitive, and trichotomy properties (as in Proposition 5.8).

However, note that in this order we have $2 > 0$ and $3 > 0$ but $2 + 3 \not> 0$. Therefore this is *not* an ordered field.

5.9 Consider the rational numbers **1** and **2** (defined as $\mathbf{1 + 1}$, which equals $[\![(2, 1)]\!]$). Of course we want $\mathbf{1} \leq \mathbf{2}$ to be true and with $a = 1, b = 1, c = 2$, and $d = 1$ this works: $1 = ad \leq bc = 2$.

However, if we take $(a, b) = (-1, -1)$, which is a different representative of **1**, we find that $-1 = ad > bc = -2$.

Hence the condition depends on the choice of representatives for **x** and **y**, giving contradictory results as to whether $\mathbf{x} \leq \mathbf{y}$ is true or not.

5.10 Let $\mathbf{x} = [\![(a, b)]\!]$, $\mathbf{y} = [\![(c, d)]\!]$, and $\mathbf{z} = [\![(e, f)]\!]$ be rational numbers.

- Reflexive: $\mathbf{x} \leq \mathbf{x}$.

    This is true because

    $$\mathbf{x} - \mathbf{x} = [\![(a, b)]\!] - [\![(a, b)]\!] = [\![(a, b)]\!] + [\![(-a, b)]\!]$$
    $$= [\![(ab - ba, b^2)]\!] = [\![(0, b^2)]\!] = \mathbf{0}.$$

- Antisymmetric: If $\mathbf{x} \leq \mathbf{y}$ and $\mathbf{y} \leq \mathbf{x}$, then $\mathbf{x} = \mathbf{y}$.

    We have $\mathbf{x} - \mathbf{y}$ and $\mathbf{y} - \mathbf{x}$ are both zero or positive. Since

    $$\mathbf{x} - \mathbf{y} = [\![(a, b)]\!] - [\![(c, d)]\!] = [\![(ad - bc, bd)]\!]$$
    $$\text{and} \quad \mathbf{y} - \mathbf{x} = [\![(c, d)]\!] - [\![(a, b)]\!] = [\![(bc - ad, bd)]\!],$$

    we see that both $ad - bc$ and $bc - ad$ are either zero or the same sign as bd. However, $ad - bc$ and $bc - ad$ are negatives of each other, so this is only possible if they are both zero. Therefore $\mathbf{x} - \mathbf{y} = \mathbf{0}$, from which it follows that $\mathbf{x} = \mathbf{y}$.

- Transitive: If $x \le y$ and $y \le z$, then $x \le z$.

  Note that if either $x = y$ or $y = z$, then the conclusion follows trivially from $x \le y \le z$.

  Therefore we can focus our attention on the case $x < y < z$. This means that $y - x$ is positive and $z - y$ is positive. By Proposition 5.9 their sum is positive, that is, $(y - x) + (z - y) = z - x > 0$ as required. [See the answer to Exercise 5.11.]

- Trichotomy: Exactly one of the following is true: $x < y$, $x = y$, or $x > y$.

  Note that $x - y = [\![(ad - bc, bd)]\!]$. If $ad - bc$ is zero, then $x = y$. If $ad - bc$ has the same sign as bd (which cannot be zero) then $x > y$. And lastly, if $ad - bc$ and bd have different signs, then $x < y$.

5.11 Let x and y be positive rational numbers. Let (a, b) be a representative of x and let (c, d) be a representative of y. Since x and y are positive, we may assume that $a, b, c,$ and d are all positive.

We have $x + y = [\![(ad + bc, bd)]\!]$ and since $ad + bc$ and bd are positive, we conclude that $x + y$ is positive.

Similarly, we have $x \cdot y = [\![(ac, bd)]\!]$ and since ac and bd are positive, we know that $x \cdot y$ is positive.

5.12 Let $z = (x + y)/2$.

5.13 Yes, $\mathbb{Z}(x)$ is a field. All of the operations behave as expected. The rational function $p(x)/q(x)$ (if different from zero) has the multiplicative inverse $q(x)/p(x)$.

Chapter 6

6.1 The left rays are $\{x \in \mathbb{Q}: x < 5\}$ and $\{x \in \mathbb{Q}: x \le 5\}$.

6.2 Claim: **L is nonempty**. Because neither L_1 nor L_2 is empty, we can find $x \in L_1$ and $y \in L_2$, and so $x + y \in L$. Therefore $L \ne \emptyset$.

Claim: **L is a proper subset of** $\mathbb{Q}$. Because both L_1 and L_2 are proper subsets of $\mathbb{Q}$, there are rational numbers $a \notin L_1$ and $b \notin L_2$. Note that all elements of L_1 are less than a and all elements of L_2 are less than b. We claim that $a + b \notin L$ because, if so, $a + b = x + y$ where $x \in L_1$ and $y \in L_2$. But $x < a$ and $y < b$ and so $x + y < a + b$, a contradiction.

Claim: **If $a < b$ and $b \in L$ then $a \in L$**. The condition $b \in L$ means $b = x + y$ where $x \in L_1$ and $y \in L_2$. Then

$$a = b + (a - b) = (x + y) + (a - b) = (x + a - b) + y.$$

Note that $a - b$ is negative (because $a < b$) and so $x + a - b < x$. Therefore $x \in L_1$. Letting $x' = x + (a - b)$ we have $a = x' + y$ where $x' \in L_1$ and $y \in L_2$. Therefore $a \in L$.

6.3 The first three left rays are all $\{x \in \mathbb{Q}: x < 2\}$ and the last one is $\{x \in \mathbb{Q}: x \le 2\}$. They are equivalent and represent the real number **2**.

Answers to Excerises 189

6.4 The decimal expansion of **−3/2** is −1.500000.... Using the method illustrated on page 87 we would have this ray:

$$L = \{x \in \mathbb{Q}: x \le -1\} \cup \{x \in \mathbb{Q}: x \le -1.5\} \cup \{x \in \mathbb{Q}: x \le -1.50\} \cup \cdots$$

Notice that additional terms in the union do not change the result because −1.5 = −1.50 = −1.500 = −1.5000 and so forth.

Further, the first term includes all rationals up to −1, so the second term doesn't add anything else. In other words, $L = \{x \in \mathbb{Q}: x \le -1\}$, which is incorrect.

One way to repair this is to first create a left ray corresponding to the positive number **3/2** and then form the additive inverse as described in Section 6.3.

6.5 Since $\pi = 3.14159265\ldots$ we can create an open left ray as the union

$$L = \{x \in \mathbb{Q}: x \le 3\}$$
$$\cup \{x \in \mathbb{Q}: x \le 3.1\}$$
$$\cup \{x \in \mathbb{Q}: x \le 3.14\}$$
$$\cup \{x \in \mathbb{Q}: x \le 3.141\}$$
$$\cup \{x \in \mathbb{Q}: x \le 3.1415\}$$
$$\cup \{x \in \mathbb{Q}: x \le 3.14159\} \cup \cdots .$$

6.6 Let $L = \{x \in \mathbb{Q}: x < 1\}$. Then a flip-swap of L gives this left ray, $\{x \in \mathbb{Q}: x \le -1\}$.

However, if L is a left ray representing $\sqrt{2}$, then a flip-swap of L will be an open left ray representing $-\sqrt{2}$.

6.7 Let L be a left ray representing **0**, say $L = \{x \in \mathbb{Q}: x < 0\}$. This is the set of negative rational numbers.

We flip this to form the right ray $\{-x \in \mathbb{Q}: x < 0\} = \{x \in \mathbb{Q}: x > 0\}$. This is the set of positive rational numbers.

Then we swap the elements of this set with the other elements of $\mathbb{Q}$. That is, we form the set $\mathbb{Q} - \{x \in \mathbb{Q}: x > 0\}$, which equals the closed left ray $\{x \in \mathbb{Q}: x \le 0\}$. This is another representative for **0**, so we have $-\mathbf{0} = \mathbf{0}$.

6.8 This is an open left ray that represents **1**: $L = \{x \in \mathbb{Q}: x < 1\}$. Note that $x \notin L$ means $x \ge 1$. Using the construction in Definition 6.14, we get this set:

$$\{x \in \mathbb{Q}: x \le 0\} \cup \{x^{-1}: x \in \mathbb{Q}, x > 0, x \notin L\}$$
$$= \{x \in \mathbb{Q}: x \le 0\} \cup \{x^{-1}: x \in \mathbb{Q}, x > 0, x \ge 1\}$$
$$= \{x \in \mathbb{Q}: x \le 0\} \cup \{x \in \mathbb{Q}: x > 0, x \le 1\}$$
$$= \{x \in \mathbb{Q}: x \le 1\},$$

which is a (closed) left ray representing **1**.

6.9 Note that for rational x if $x \notin L$ then $x \ge 2$. Therefore

$$\{x \in \mathbb{Q}: x \le 0\} \cup \{x^{-1}: x \in \mathbb{Q}, x > 0, x \notin L\} = \{x \in \mathbb{Q}: x \le \tfrac{1}{2}\}.$$

Similarly, if $x \notin L'$ then $x > 2$. Therefore

$$\{x \in \mathbb{Q}: x \le 0\} \cup \{x^{-1}: x \in \mathbb{Q}, x > 0, x \notin L'\} = \{x \in \mathbb{Q}: x < \tfrac{1}{2}\}.$$

These are equivalent left rays representing the real number $\tfrac{1}{2}$.

6.10 If x is an even integer, then $x = 2a$ for some integer a. Then $x^2 = (2a)^2 = 4a = 2 \cdot (2a)$ and is therefore even.

On the other hand, if x is an odd integer, then $x = 2a + 1$ for some integer a. Then $x^2 = (2a + 1)^2 = 4a^2 + 4a + 1 = 2 \cdot (2a + 1) + 1$, which is also odd.

6.11
- Suppose, for the sake of contradiction that $\sqrt{2} = b/a$ where b and a are positive integers that are as small as possible.
- If a and b have a common divisor d, then $a' = a/d$ and $b' = b/d$ are integers. Notice that
$$\left(\frac{b'}{a'}\right)^2 = \left(\frac{b/d}{a/d}\right)^2 = \frac{b^2}{a^2} = 2,$$
showing that $b'/a' = \sqrt{2}$ but a' and b' are smaller than a and b. However, a and b were the smallest possible positive integers with $b/a = \sqrt{2}$. Hence a and b do not have a common divisor.
- From $(b/a)^2 = 2$ we have $b^2 = 2a^2$. Therefore b^2 is even. By Exercise 6.10 we know b isn't odd (otherwise b^2 would be odd) and so b is even.

Therefore $b = 2x$ for some integer x.
- From $b^2 = 2a^2$ we have $(2x)^2 = 2a^2$, which gives $4x^2 = 2a^2$ and so $2x^2 = a^2$. This implies a^2 is even and, as we argued in the previous step, that implies a is even.
- We have shown that a and b are both even, so they have a common divisor of 2, but we argued that a and b have no common divisor, and this is a contradiction.
- Therefore there are no integers a and b such that $b/a = \sqrt{2}$.

6.12 Suppose, for the sake of contradiction, there is a rational number $x = a/b$ (where $a, b \in \mathbb{Z}$) such that $2^x = 3$. In other words, $2^{a/b} = 3$, which implies $2^a = 3^b$.

Note that we may assume that both a and b are positive, or exactly one of them is positive and the other is negative. (Were a and b both less than zero, we could replace them by their negatives.)

If a and b are both positive, then we have 2^a (which is an even number) equal to 3^b (which is an odd number), and that's impossible.

If only one of a or b is positive (and the other is negative), then one of 2^a or 3^b is less than 1 and the other is greater than 1, and that's also impossible.

In either case, we reach an impossible conclusion. Therefore the supposition that $\log_2 3$ is a rational number is false.

6.13 Let **x** and **y** be positive rational numbers represented by left rays L and R, respectively.

By Proposition 6.10 we can find positive rational numbers $\ell \in L$ and $r \in R$. Then $L + R$ (which represents **x** + **y**) contains $\ell + r$, which is positive.

Therefore, by Proposition 6.10, **x** + **y** is positive.

6.14 The set $\{x \cdot y : x \in L_2 \text{ and } y \in L_3\}$ is all $\mathbb{Q}$.

6.15 a) Let $L = \{x \in \mathbb{Q} : x < 0\}$ be the open left ray that represents **0**, and let R be a left ray representing **x**. From Definition 6.11, the set

$$\{a \cdot b : a \in L, \ a \geq 0, \ b \in R, \ \text{and} \ b \geq 0\}$$

is the empty set because there is no element $a \in L$ such that $a \geq 0$. Therefore the ray created by Definition 6.11 is

$$\emptyset \cup \{x \in \mathbb{Q} : x < 0\},$$

which is the open ray representative of **0**.

b) Similarly, let $L' = \{x \in \mathbb{Q} : x \leq 0\}$ be the closed ray representing **0**. In this case the set

$$\{a \cdot b : a \in L', \ a \geq 0, \ b \in R, \ \text{and} \ b \geq 0\}$$

is simply $\{0\}$ because the only element of L' that is at least zero is 0. Therefore, the ray created by Definition 6.11 is

$$\{0\} \cup \{x \in \mathbb{Q} : x < 0\} = \{x \in \mathbb{Q} : x \leq 0\},$$

which is the closed ray representative of **0**.

In either case, we see that $\mathbf{0} \cdot \mathbf{x} = \mathbf{0}$, as expected.

6.16 a) The left rays are of the form $\{x \in \mathbb{Q} : x < -1/n\}$ where n is a positive integer. Alternatively we can use closed rays, $\{x \in \mathbb{Q} : x \leq -1/n\}$.

b) The union of these rays is the left ray $R = \{x \in \mathbb{Q} : x < 0\}$. It is an open ray. Note that 0 cannot possibly be in R because 0 is not an element of any of the sets L_x.

c) The real number represented by R is **0**. It's clear that **0** is an upper bound for X because all the elements of X are negative. Further, it's not possible for a negative number, say **y**, to be an upper bound for X because for **n** sufficiently large we would have $\mathbf{y} < -\mathbf{1/n}$. Therefore **y** is not an upper bound for X.

6.17 To begin, L consists only of negative rational numbers; that's implied by the $a < 0$ condition. The added condition is that elements a of L must satisfy $a^2 \geq 2$. For example, -3 satisfies this condition (because $(-3)^2 = 9 > 2$) but -1 does not. So some negative numbers are ruled out by this condition.

However, suppose a is in L and $b < a$. We have to be sure that b is also in L. Note that $b^2 > a^2$ (the inequality flips because a and b are negative) and we know that $a^2 \geq 2$. Therefore $b^2 \geq 2$ and so $b \in L$.

All this verifies that L is a left ray.

Finally, we see that this ray represents $-\sqrt{2}$.

6.18 Suppose **a** and **b** are least upper bounds for a set X. Then $\mathbf{a} \leq \mathbf{b}$ (because **a** is a least upper bound) and $\mathbf{b} \leq \mathbf{a}$ (because **b** is a least upper bound). Therefore $\mathbf{a} = \mathbf{b}$.

6.19 Let $X = \{x \in \mathbb{Q} : x^2 < 2\}$. Then X has an upper bound (e.g., all elements of X are less than or equal to 2) but it does not have a least upper bound because there is no $\sqrt{2}$ in $\mathbb{Q}$.

Chapter 7

7.1 If x and y are both nonnegative, then $|x| + |y| = x + y = |x + y|$. If x and y are both nonpositive, then $|x| + |y| = (-x) + (-y) = -(x + y) = |x + y|$.

The interesting case is when one term is positive and the other is negative. By symmetry, say x is positive and y is negative.

- If $|x| > |y|$ then $x + y$ is positive, but less than x. Therefore $|x + y| < |x| < |x| + |y|$.
- If $|x| < |y|$ then $x+y$ is negative, but not as negative as y; that is, $y < x+y < 0$. Therefore $|x + y| = -(x + y) < -y = |y|$.
- Finally, if $|x| = |y|$, then $x + y = 0$ and so $|x + y| = 0 \leq |x| + |y|$.

In every case, $|x + y| \leq |x| + |y|$.

7.2 Using the triangle inequality: $|x - y| = |x + (-y)| \leq |x| + |-y| = |x| + |y|$.

7.3 Let s be a "small" positive number. We have to show that, "eventually," all differences $|a_i - a_j| < s$.

We have
$$|a_i - a_j| = \left|\left(1 + \frac{(-1)^i}{i}\right) - \left(1 + \frac{(-1)^j}{j}\right)\right|$$
$$= \left|\frac{(-1)^i}{i} - \frac{(-1)^j}{j}\right| \leq \left|\frac{(-1)^i}{i}\right| + \left|\frac{(-1)^j}{j}\right| = \frac{1}{i} + \frac{1}{j}.$$

Pick a positive rational number s and choose N such that $N > 2/s$. If $i, j > N$ we have
$$|a_i - a_j| \leq \frac{1}{i} + \frac{1}{j} < \frac{1}{N} + \frac{1}{N} < s.$$

This shows that $\hat{a}$ is Cauchy.

Next we show that $\hat{a} \equiv (1, 1, 1, \ldots)$. Choose $s > 0$. We need to show that eventually $|a_j - 1| < s$. Take $N > 1/s$, then we have
$$|a_j - 1| = \left|\left(1 + \frac{(-1)^j}{j}\right) - 1\right| = \frac{1}{j} < \frac{1}{N} < s.$$

Therefore $\hat{a} \equiv (1, 1, 1, \ldots)$.

7.4 To show that $\hat{a} \equiv \hat{b}$, we pick any positive rational number s and we have to show that, eventually, the differences between the corresponding entries in $\hat{a}$ and $\hat{b}$ are all less than s.

We know that $\hat{a}$ is Cauchy, so there is a number N such that for all $i, j \geq N$ we have $|a_i - a_j| < s$. In particular, if we take $j = i + 1$, we have $|a_i - a_{i+1}| < s$. Note that a_{i+1} is simply the ith term of $\hat{b}$. Hence we have shown that $\hat{a} \equiv \hat{b}$.

Answers to Excerises 193

Next we have $\hat{c}$ composed of the even indexed terms of $\hat{a}$. We have to show that eventually all the differences between corresponding terms of $\hat{a}$ and $\hat{c}$ are less than s. That is, $|a_i - c_i| < s$ once $i > N$. But note that $c_i = a_{2i}$ and we know that $|a_i - a_j| < s$ for all $i, j > N$. Since $2i > i$, we have $|a_i - c_i| = |a_i - a_{2i}| < s$ and therefore $\hat{a} \equiv \hat{c}$.

7.5 We know that eventually all differences $|a_i - a_j| < 1$. Suppose that happens by index N, that is, for all $i, j \geq N$ we have $|a_i - a_j| < 1$. In particular, $|a_N - a_j| < 1$ for all $j \geq N$. This means that a_j is less than 1 away from a_N and that implies that $|a_j| < |a_N| + 1$.

We have established that for indices $j \geq N$ we have $|a_j| < |a_N| + 1$.

Let $B = \max\{|a_1|, |a_2|, |a_3|, \ldots, |a_N|\} + 1$. It follows that $|a_j| < B$ for all j (regardless of whether $j \geq N$ or $1 \leq j \leq N$).

7.6 Suppose that $\hat{a}$ is bounded and increasing, but (for sake of contradiction) is not Cauchy. This means there is a positive gap size s so that at no point are all differences at most s.

This means that we can find indices $i_1 < i_2 < i_3 < i_4 < \cdots$ so that

$$a_{i_2} - a_{i_1} > s,$$
$$a_{i_4} - a_{i_3} > s,$$
$$a_{i_6} - a_{i_5} > s,$$
$$a_{i_8} - a_{i_7} > s,$$

and so on.

Because the sequence is increasing, we have

$$a_{i_2} > a_{i_1} + s,$$
$$a_{i_4} > a_{i_3} + s > a_{i_2} + s > (a_{i_1} + s) + s = a_{i_1} + 2s,$$
$$a_{i_6} > a_{i_5} + s > a_{i_4} + s > (a_{i_1} + 2s) + s = a_{i+1} + 3s,$$
$$a_{i_8} > a_{i_7} + s > a_{i_6} + s > (a_{i_1} + 3s) + s = a_{i+1} + 4s,$$

and so on. In other words, $a_{i_{2k}} > a_{i_1} + ks$. Since $s > 0$ and we can take k as large as we like, we see that $\hat{a}$ is not bounded. This is a contradiction, hence our supposition that $\hat{a}$ is not Cauchy is not true. Therefore $\hat{a}$ is Cauchy.

7.7 Let s be a "small" positive rational number. We have to show that eventually $|a_i b_i - a_j b_j| < s$.

We know from Exercise 7.5 that the sequences $\hat{a}$ and $\hat{b}$ are bounded. If $\hat{a}$ is bounded by A and $\hat{b}$ is bounded by B, then $X = \max\{A, B\}$. Hence, all elements of the two sequence are bounded by X.

Now consider the differences $a_i b_i - a_j b_j$; we need show these differences are small. We do the following algebra:

$$|a_i b_i - a_j b_j| = |(a_i - a_j)b_i + (b_i - b_j)a_j|$$
$$\leq |b_i| \cdot |a_i - a_j| + |a_j| \cdot |b_i - b_j|$$
$$\leq X|a_i - a_j| + X|b_i - b_j| = X\left(|a_i - a_j| + |b_i - b_j|\right).$$

Since $\hat{a}$ and $\hat{b}$ are Cauchy, for any positive rational t, there is an index N beyond which all the differences $|a_i - a_j|$ and $|b_i - b_j|$ are less than t. To ensure that $|a_i b_i - a_j b_j| < s$, we take $t = s/(2X)$.

Continuing the calculation from above:

$$|a_i b_i - a_j b_j| \le X\left(|a_i - a_j| + |b_i - b_j|\right)$$
$$< X(t+t) = X\left(\frac{s}{2X} + \frac{s}{2X}\right) = s.$$

7.8 This is not a Cauchy sequence. Let's estimate the difference between terms h_n and h_{2n}. We have:

$$h_{2n} = 1 + \tfrac{1}{2} + \tfrac{1}{3} + \cdots + \tfrac{1}{n} + \tfrac{1}{n+1} + \cdots + \tfrac{1}{2n},$$
$$h_n = 1 + \tfrac{1}{2} + \tfrac{1}{3} + \cdots + \tfrac{1}{n}$$
$$\Rightarrow h_{2n} - h_n = \tfrac{1}{n+1} + \cdots + \tfrac{1}{2n}.$$

Note that all of the n terms in the expression for $h_{2n} - h_n$ are at least $1/(2n)$. Therefore

$$h_{2n} - h_n = \tfrac{1}{n+1} + \cdots + \tfrac{1}{2n} \ge \underbrace{\tfrac{1}{2n} + \cdots + \tfrac{1}{2n}}_{n \text{ terms}} = \tfrac{1}{2}.$$

If we take $s = \tfrac{1}{3}$, then there is no point in $\hat{h}$ after which *all* differences are less than $\tfrac{1}{2}$. Therefore $\hat{h}$ is not a Cauchy sequence, even though the difference between *consecutive* terms gets smaller and smaller.

7.9 To show that $\hat{a} \equiv (0,0,0,\ldots)$, pick a rational number $s > 0$. We have to show that eventually $|a_i| = |a_i - 0| < s$. To this end, pick a large integer N such that $|a_i - a_j| < s$ for all $i, j \ge N$. Since there are infinitely many elements of $\hat{a}$ equal to 0, there is a $j > N$ with $a_j = 0$. Since $|a_i - a_j| < s$ (because $i, j > N$) and since $a_j = 0$, we conclude $|a_i| < s$.

7.10 We need to show that if $s > 0$ we can find an N such that $|a_i - 0| = |a_i| < s$ for all $i \ge N$.

Since $\hat{a}$ is Cauchy, we can choose N so that for all $i, j \ge N$ we have $|a_i - a_j| < s$.

If a_i is positive, find an a_j (with $j > i$) so that a_j is negative. This is possible because $\hat{a}$ has infinitely many negative values. It now follows that $|a_i| < |a_i - a_j| < s$.

Likewise, if a_i is negative, find an a_j (with $j > i$) so that a_j is positive. It now follows that $|a_i| < |a_i - a_j| < s$.

Therefore $\hat{a} \equiv (0,0,0,\ldots)$.

7.11 Let $\hat{x}$ be a representative of $\mathbf{x}$ with $\mathbf{x} \ne \mathbf{0}$. If $\hat{x}$ had infinitely many 0s then, by Exercise 7.9, $\hat{x} \equiv (0,0,0,\ldots)$, but that contradicts $\mathbf{x} \ne \mathbf{0}$.

7.12 We know that $\hat{a}$ is positive so for some positive rational number s and a positive integer N_1 we have $a_n > s$ for all $n > N_1$.

We also know that $\hat{b} \equiv \hat{a}$. That means we can find an integer N_2 so that for all $n > N_2$ we have $|b_n - a_n| < s/2$.

Let N be the larger of N_1 and N_2. For $n > N$ we have
$$b_n = a_n + (b_n - a_n) > s + (b_n - a_n) > s - \frac{s}{2} = \frac{s}{2},$$
and therefore $\hat{b}$ is positive.

7.13 Consider the real number $\mathbf{x} - \mathbf{y}$ which is represented by the sequence
$$(a_1 - b_1, a_2 - b_2, a_3 - b_3, \ldots).$$
Notice that all the terms in this sequence are either negative or zero. Therefore $\mathbf{x} - \mathbf{y}$ cannot be positive. Therefore $\mathbf{x} - \mathbf{y} \leq \mathbf{0}$ or $\mathbf{x} \leq \mathbf{y}$.

7.14 Let $\hat{a}$ be a representative of $\mathbf{a}$. Since $\mathbf{a}$ is nonzero, we know that $\hat{a}$ has only finitely many 0s (Exercise 7.11). Further, it cannot have infinitely many positive and negative values (Exercise 7.10).

Suppose first that $\hat{a}$ has only finitely many negative values. That means eventually all entries in $\hat{a}$ are positive. This implies that $\mathbf{a} \geq \mathbf{0}$, but since $\mathbf{a}$ is nonzero, we have $\mathbf{a}$ is positive.

On the other hand, if $\hat{a}$ has only finitely many positive values, then a representative of $-\mathbf{a}$ has only finitely many negative values, and so $-\mathbf{a}$ is positive.

7.15 Choose representatives $\hat{a}$ and $\hat{b}$ for $\mathbf{a}$ and $\mathbf{b}$. Since these numbers are positive, there are positive rationals s and t such that, eventually, all elements of $\hat{a}$ are greater than s and, eventually, all elements of $\hat{b}$ are greater then t.

It follows that, eventually, all elements of $\hat{a} + \hat{b}$ are greater than $s + t$ (which is positive) and all elements of $\hat{a} \cdot \hat{b}$ are greater than st. Therefore $\mathbf{a} + \mathbf{b}$ and $\mathbf{a} \cdot \mathbf{b}$ are positive.

7.16 a) The values of the first several terms in $\hat{a}$, written in binary, are:

1.
1.0
1.01
1.011
1.0110
1.01101
1.011010
1.0110101
1.01101010
1.011010100
1.0110101000
1.01101010000
1.011010100000
1.0110101000001
1.01101010000010
1.011010100000100
1.0110101000001001
1.01101010000010011
1.011010100000100111
1.0110101000001001111

b) In binary, $\sqrt{2}$ is 1.01101010000010011111....

c) The successive terms in $\hat{a}$ add one binary digit from the base-two representation of $\sqrt{2}$.

7.17 Here is a table showing the iterations for the two methods.

Step	Bisection midpoint	Newton x
0	1.25	1.5
1	1.375	1.4166666666666667
2	1.4375	1.4142156862745099
3	1.40625	1.4142135623746899
4	1.421875	1.4142135623730951
5	1.4140625	1.414213562373095

We compare the results after five iterations:

Method	Final x	x^2	Error ($x^2 - 2$)
Bisection	1.4140625	1.99957275390625	-4.27×10^{-4}
Newton	1.414213562373095	1.9999999999999996	-4.44×10^{-16}

Newton's method is vastly superior.

Chapter 8

8.1 a) Observe that for $a + b\sqrt{2}$ and $c + d\sqrt{2}$, elements of $\mathbb{Q}\left[\sqrt{2}\right]$, we have

$$(a + b\sqrt{2}) + (c + d)\sqrt{2} = (a + c) + (b + d)\sqrt{2},$$

and therefore their sum is also in $\mathbb{Q}\left[\sqrt{2}\right]$.

Similarly,

$$(a + b\sqrt{2}) \cdot (c + d\sqrt{2}) = (ac + 2bd) + (ad + bc)\sqrt{2},$$

and therefore their product is in $\mathbb{Q}\left[\sqrt{2}\right]$.

b) For $x = a + b\sqrt{2} \in \mathbb{Q}\left[\sqrt{2}\right]$ let

$$y = \left(\frac{a}{a^2 - 2b^2}\right) - \left(\frac{b}{a^2 - 2b^2}\right)\sqrt{2}.$$

Notice that as long as one of a or b is nonzero, the denominators are not zero (because we can't have $a^2 = 2b^2$ because $\sqrt{2} \notin \mathbb{Q}$).

Now we just multiply them together:

$$x \cdot y = \left[a + b\sqrt{2}\right] \cdot \left[\left(\frac{a}{a^2 - 2b^2}\right) - \left(\frac{b}{a^2 - 2b^2}\right)\sqrt{2}\right]$$

$$= \left(\frac{a^2 - 2b^2}{a^2 - 2b^2}\right) + \left(\frac{-ab + ba}{a^2 - 2b^2}\right)\sqrt{2}$$

$$= 1.$$

c) Note that if both a and b are negative or zero, then $a + b\sqrt{2}$ is not positive (it's either negative or zero).

If $a \geq 0$ and $b \geq 0$, and they are not both zero, then $a + b\sqrt{2}$ is positive.

The interesting cases arise when one of a or b is positive and the other is negative.

If we have $a > 0$ and $b < 0$ then determining if $a + b\sqrt{2} > 0$ is the same as determining if $a > (-b)\sqrt{2}$ (both sides are positive numbers). That happens exactly when $a^2 > 2b^2$.

If $a < 0$ and $b > 0$ then determining if $a + b\sqrt{2} > 0$ is the same as determining if $b\sqrt{2} > -a$ (both sides are positive numbers). That happens exactly when $2b^2 > a^2$.

Summarizing, assuming a and b are not both zero, we have the following chart:

Condition on a	Condition on b	Is $a + b\sqrt{2}$ positive?
$a \geq 0$	$b \geq 0$	Yes
$a \leq 0$	$b \leq 0$	No
$a > 0$	$b < 0$	Yes, when $a^2 > 2b^2$
$a < 0$	$b > 0$	Yes, when $a^2 < 2b^2$

8.2 Here are the powers of 2 in $\mathbb{Z}_{11}^*$:

k	0	1	2	3	4	5	6	7	8	9	10
2^k	1	2	4	8	5	10	9	7	3	6	1

If we associate a in $\mathbb{Z}_{10}$ with 2^a in $\mathbb{Z}_{11}^*$ we have a one-to-one correspondence and notice that $a + b$ in $\mathbb{Z}_{10}$ corresponds to $2^a \cdot 2^b$ in $\mathbb{Z}_{11}^*$ because $2^a \cdot 2^b = 2^{a+b}$ (in $\mathbb{Z}_{11}^*$).

Now consider the powers of 3 in $\mathbb{Z}_{11}^*$:

k	0	1	2	3	4	5	6	7	8	9	10
3^k	1	3	9	5	4	1	3	9	5	4	1

This does not give a one-to-one correspondence because (for example) both 2 and 7 in $\mathbb{Z}_{11}^*$ are paired with 9 in $\mathbb{Z}_{10}$, and (for example) there is no element of $\mathbb{Z}_{11}^*$ paired with 8 in $\mathbb{Z}_{10}$.

8.3 We start with $f(1) = 3$ and we require that $f(a+b) = f(a) + f(b)$.

We know that $f(1) = 3$. What is $f(2)$? Since $2 = 1 + 1$ we have $f(2) =$

$f(1) + f(1) = 3 + 3 = 6$. Continuing this way, we build up as follows:

$$f(1) = 3$$
$$f(2) = f(1 + 1) = f(1) + f(1) = 3 + 3 = 6,$$
$$f(3) = f(2 + 1) = f(2) + f(1) = 6 + 3 = 9,$$
$$f(4) = f(3 + 1) = f(3) + f(1) = 9 + 3 = 2,$$
$$f(5) = f(4 + 1) = f(4) + f(1) = 2 + 3 = 5,$$
$$f(6) = f(5 + 1) = f(5) + f(1) = 5 + 3 = 8,$$
$$f(7) = f(6 + 1) = f(6) + f(1) = 8 + 3 = 1,$$
$$f(8) = f(7 + 1) = f(7) + f(1) = 1 + 3 = 4,$$
$$f(9) = f(8 + 1) = f(8) + f(1) = 4 + 3 = 7,$$
$$f(0) = f(9 + 1) = f(9) + f(1) = 7 + 3 = 0.$$

Writing out the addition tables in the two forms (merged together) shows that we always have $f(a + b) = f(a) + f(b)$:

+	0(0)	1(3)	2(6)	3(9)	4(2)	5(5)	6(8)	7(1)	8(4)	9(7)
0(0)	0(0)	1(3)	2(6)	3(9)	4(2)	5(5)	6(8)	7(1)	8(4)	9(7)
1(3)	1(3)	2(6)	3(9)	4(2)	5(5)	6(8)	7(1)	8(4)	9(7)	0(0)
2(6)	2(6)	3(9)	4(2)	5(5)	6(8)	7(1)	8(4)	9(7)	0(0)	1(3)
3(9)	3(9)	4(2)	5(5)	6(8)	7(1)	8(4)	9(7)	0(0)	1(3)	2(6)
4(2)	4(2)	5(5)	6(8)	7(1)	8(4)	9(7)	0(0)	1(3)	2(6)	3(9)
5(5)	5(5)	6(8)	7(1)	8(4)	9(7)	0(0)	1(3)	2(6)	3(9)	4(2)
6(8)	6(8)	7(1)	8(4)	9(7)	0(0)	1(3)	2(6)	3(9)	4(2)	5(5)
7(1)	7(1)	8(4)	9(7)	0(0)	1(3)	2(6)	3(9)	4(2)	5(5)	6(8)
8(4)	8(4)	9(7)	0(0)	1(3)	2(6)	3(9)	4(2)	5(5)	6(8)	7(1)
9(7)	9(7)	0(0)	1(3)	2(6)	3(9)	4(2)	5(5)	6(8)	7(1)	8(4)

However, if we begin with $g(1) = 2$ we run into trouble. As before:

$$g(1) = 2,$$
$$g(2) = g(1 + 1) = g(1) + g(1) = 2 + 2 = 4,$$
$$g(3) = g(2 + 1) = g(2) + g(1) = 4 + 2 = 6,$$
$$g(4) = g(3 + 1) = g(3) + g(1) = 6 + 2 = 8,$$
$$g(5) = g(4 + 1) = g(4) + g(1) = 8 + 2 = 0,$$
$$g(6) = g(5 + 1) = g(5) + g(1) = 0 + 2 = 2,$$
$$g(7) = g(6 + 1) = g(6) + g(1) = 2 + 2 = 4,$$
$$g(8) = g(7 + 1) = g(7) + g(1) = 4 + 2 = 6,$$
$$g(9) = g(8 + 1) = g(8) + g(1) = 6 + 2 = 8,$$
$$g(0) = g(9 + 1) = g(9) + g(1) = 8 + 2 = 0.$$

The problem is that we do not have a one-to-one correspondence.

8.4 Consider terms a_n and a_m of this sequence. The difference between these terms is

$$|a_n - a_m| = \left| \frac{1}{\sqrt{n}} - \frac{1}{\sqrt{m}} \right| \leq \frac{1}{\sqrt{n}} + \frac{1}{\sqrt{m}}$$

by the triangle inequality (see Proposition 7.3).

If $n, m \geq N$ (for some large positive integer N), then we have $|a_n - a_m| \leq \frac{2}{\sqrt{N}}$.

If follows that if we want $|a_n - a_m| < s$ (for some "small" positive number s), we just need to take $N > 4s^2$.

8.5 Sorry Mr. Bond, but 007 is not a valid representation of the natural number 7.

The following rules describe when a finite sequence $s_1 s_2 s_3 \ldots s_n$ is a natural numbers:

- If the sequence has only one symbol, s_1, then s_1 may be any one of the ten digits 0 through 9.

- Otherwise, the first symbol, s_1, may be any one of the nine digits 1 through 9, and the remaining symbols may be any of the ten digits 0 through 9.

8.6 Let X be an integer.

- If $X \geq 0$ (so X is a natural number) then the decimal notation for X is exactly the same as the decimal notation for X as a natural number (as laid out in Exercise 8.5).

- Otherwise, if X is negative, let $d_1 d_2 d_3 \ldots d_n$ be the decimal representation of $-X$. Then the decimal representation of X is $s d_1 d_2 \ldots d_n$ where s is $-$ (the minus sign).

8.7 A decimal representation of a real number starts with a decimal representation of an integer (see Exercise 8.6).

This is optionally followed by a decimal point. If there is no decimal point, there are no further symbols.

If there is a decimal point, then it is followed by a finite or infinite stream of digits.

However, it is permissible to begin a base-ten representation with a decimal point (or a minus sign and a decimal point), as long as there is at least one digit following it. That is, .25 is a valid[6] decimal representation of the real number $\frac{1}{4}$ and $-.25$ is $-\frac{1}{4}$.

8.8 One way to tackle this is to define a *standard form* for decimal representations. Once the standard form is established, two decimal representations are equivalent (are the same real number) exactly when they have the same standard form.

Suppose x is written in decimal. Here is how we transform this representation into a standard form.

First, if there is no decimal point (so the number is an integer) append a decimal point. For example, $123 \to 123.$

Second, if there are no digits to the left of the decimal point, insert a 0 just to the left of the decimal point. For example, $-.123 \to -0.123$.

Third, if the decimal representation ends with an infinite stream of 9s, change those 9s to 0s, and increase the rightmost digit that isn't a 9 by 1. For example, $2.4999999\ldots \to 2.5000000\ldots$ or $-29.99999\ldots \to -30.00000\ldots$. However, if the representation digits are all 9s, then transform all the 9s to 0s, and insert a 1 at the left. For example, $99.9999\ldots \to 100.0000\ldots$

---

[6]The equivalent representation 0.25 is preferred to make the decimal point more noticeable. However, this is a typographical convention, not a mathematical requirement.

Finally, if the decimal representation consists of only finitely many digits, append an infinite list of 0s to the end.

8.9 Scientific notation for real numbers is written in this form:
$$A \times 10^N$$
where A is a decimal number with $1 \leq |A| < 10$ and N is a decimal integer. In the special case $A = 1$, then A should not be written as $0.999999\ldots$ (and likewise for $A = -1$).

In practice, A is written only with finitely many digits as scientific notation is used for actual measurements and those do not have infinite precision.

There is no way to write the number 0 given the rules we laid out. We can resolve that by allowing 0×10^0 to be its scientific notation.

8.10 Since $x = 5.3520153\ldots$ and $y = 2.498151\ldots$, we know that
$$5.352 \leq x \leq 5.353$$
$$\text{and} \quad 2.498 \leq y \leq 2.499.$$

It follows that
$$7.8\underline{50} = 5.352 + 2.498 \leq x + y \leq 5.353 + 2.499 = 7.8\underline{52},$$
and so the first two digits after the decimal point for $x + y$ are 85.

Note that the two-digit bounds on x and y are not sufficient to determine the first two digits after the decimal point for xy:
$$13.\underline{36}9296 = 5.352 \times 2.498 \leq xy \leq 5.353.49913.\underline{37}7147,$$

Using four-digit bounds is also insufficient. However, if we consider five-digit bounds we find
$$13.\underline{37}01237815 = 5.35201 \times 2.49815 \leq xy \leq 5.35202 \times 2.49816 = 13.\underline{37}02022832,$$
which shows that the first two digits after the decimal point are 37.

Chapter 9

9.1 Let $w = a + bi$ and $z = c + di$. Then
$$\begin{aligned}
|w \cdot z| &= |(a + bi) \cdot (c + di)| \\
&= |(ac - bd) + (ad + bc)i| \\
&= \sqrt{(ac - bd)^2 + (ad + bc)^2} \\
&= \sqrt{a^2c^2 + b^2c^2 + a^2d^2 + b^2d^2} \\
&= \sqrt{(a^2 + b^2)(c^2 + d^2)} \\
&= \sqrt{a^2 + b^2} \cdot \sqrt{c^2 + d^2} \\
&= |w| \cdot |z|.
\end{aligned}$$

9.2 Since $|z|$ is the distance from z to the origin, we have $|z| = r$.
Alternatively, we have $z = (r\cos\theta) + (r\sin\theta)i$. Therefore
$$|z| = \sqrt{r^2\cos^2\theta + r^2\sin^2\theta} = r\sqrt{\cos^2\theta + \sin^2\theta} = r\sqrt{1} = r.$$

9.3 The equation $z^3 = 1$ can be rewritten $z^3 - 1 = 0$. Factoring the left-hand side gives $(z-1)(z^2+z+1) = 0$. The additional cube roots of 1 can be found using the quadratic formula to give $\frac{1}{2}(-1+i\sqrt{3})$ and $\frac{1}{2}(-1-i\sqrt{3})$.

Alternatively, the number 1 can be written in polar coordinates in any of the following ways: $(1,0)$, $(1,2\pi)$, or $(1,4\pi)$.

Suppose z is a cube root of 1 with polar coordinates (r,θ). Then z^3 has polar coordinates $(r^3, 3\theta)$.

- If we set $(r^3, 3\theta) = (1,0)$, we find $(r,\theta) = (1,0)$, giving $z = 1$.
- If we set $(r^3, 3\theta) = (1, 2\pi)$, then we find $(r,\theta) = (1, \frac{2}{3}\pi)$, which gives
$$z = \cos(\tfrac{2}{3}\pi) + i\sin(\tfrac{2}{3}\pi) = -\tfrac{1}{2} + \tfrac{\sqrt{3}}{2}i.$$
- Similarly, if we set $(r^3, 3\theta) = (1, 4\pi)$, we find that $(r,\theta) = (1, \frac{4}{3}\pi)$, leading to $z = -\tfrac{1}{2} - \tfrac{\sqrt{3}}{2}i$.

9.4 If $(\mathbb{C}, +, \cdot, \preceq)$ were an ordered field, it would be the case that if $w \succ 0$ and $z \succ 0$ then $w \cdot z \succ 0$.

However, if we take $w = 1 + 2i \succ 0$ and $z = 1 + 3i \succ 0$, we would have to have $w \cdot z \succ 0$, but in fact $w \cdot z = -5 + 5i \prec 0$.

9.5 Let $\equiv$ stand for equivalence modulo $x^2 + 1$.

- $\equiv$ is reflexive. Let $p(x) \in \mathbb{R}[x]$. Since $p(x) - p(x) = 0$ is a multiple of $x^2 + 1$ we have $p(x) \equiv p(x)$.
- $\equiv$ is symmetric. Let $p(x), q(x) \in \mathbb{R}[x]$ and suppose $p(x) \equiv q(x)$. This means that $p(x) - q(x)$ is a multiple of $x^2 + 1$. But then $q(x) - p(x)$, which is simply $-[p(x) - q(x)]$ is also a multiple of $x^2 + 1$ and therefore $q(x) \equiv p(x)$.
- $\equiv$ is transitive. Let $p(x), q(x), r(x) \in \mathbb{R}[x]$ and suppose we have $p(x) \equiv q(x)$ and $q(x) \equiv r(x)$. This implies
$$p(x) - q(x) = a(x)\left[x^2 + 1\right] \quad \text{and}$$
$$q(x) - r(x) = b(x)\left[x^2 + 1\right]$$
where $a(x)$ and $b(x)$ are polynomials. Adding these together gives
$$p(x) - r(x) = [p(x) - q(x)] + [q(x) - r(x)] = [a(x) + b(x)]\left[x^2 + 1\right],$$
and so $p(x) \equiv r(x)$.

Therefore $\equiv$ is an equivalence relation on $\mathbb{R}[x]$.

9.6 Given that $p_1(x) \equiv p_2(x)$ and $q(x) \equiv q_2(x)$, we know that

$$p_1(x) - p_2(x) = a(x)\left(x^2 + 1\right) \quad \text{and} \quad q_1(x) - q_2(x) = b(x)\left(x^2 + 1\right)$$

for some polynomials $a(x)$ and $b(x)$. These can be rewritten as

$$p_1(x) = p_2(x) + a(x)\left(x^2 + 1\right) \quad \text{and} \quad q_1(x) = q_2(x) + b(x)\left(x^2 + 1\right). \quad (*)$$

Adding the two equations in $(*)$ gives

$$p_1(x) + q_1(x) = p_2(x) + q_2(x) + [a(x) + b(x)]\left(x^2 + 1\right),$$

and so $p_1(x) + q_1(x) \equiv p_2(x) + q_2(x)$.

Similarly, multiplying the two equations in $(*)$ gives

$$p_1(x) \cdot p_2(x) = \left[p_2(x) + a(x)\left(x^2 + 1\right)\right] \cdot \left[q_2(x) + b(x)\left(x^2 + 1\right)\right]$$

$$= p_2(x)q_2(x) + [a(x)q_2(x) + b(x)p_2(x)]\left(x^2 + 1\right)$$

$$+ a(x)b(x)\left(x^2 + 1\right)^2$$

$$= p_2(x)q_2(x)$$

$$+ \left[a(x)q_2(x) + b(x)p_2(x) + a(x)b(x)\left(x^2 + 1\right)\right]\left(x^2 + 1\right)$$

$$= p_2(x)q_2(x) + r(x)\left(x^2 + 1\right)$$

where $r(x) = a(x)q_2(x) + b(x)p_2(x) + a(x)b(x)(x^2+1)$. Therefore $p_1(x)q_1(x) \equiv p_2(x)q_2(x)$.

9.7 Polar coordinates (r_1, θ_1) and (r_2, θ_2) are equivalent provided either (a) $r_1 = r_2 \neq 0$ and $\theta_1 - \theta_2$ is an integer multiple of 2π or (b) $r_1 = r_2 = 0$.

9.8 Let $w = a + bi$ and $z = c + di$.

a) $M_w + M_z = \begin{bmatrix} a & b \\ -b & a \end{bmatrix} + \begin{bmatrix} c & d \\ -d & c \end{bmatrix} = \begin{bmatrix} a+c & b+d \\ -(b+d) & a+c \end{bmatrix} = M_{w+z}$.

b) $M_w \cdot M_z = \begin{bmatrix} a & b \\ -b & a \end{bmatrix} \cdot \begin{bmatrix} c & d \\ -d & c \end{bmatrix} = \begin{bmatrix} ac-bd & ad+bc \\ -(ad+bc) & ac-bd \end{bmatrix} = M_{wz}$.

c) $\det M_w = \det \begin{bmatrix} a & b \\ -b & a \end{bmatrix} = a^2 + b^2 = |w|^2$.

9.9 a) Here are the tables:

+	0+0i	1+0i	2+0i	0+1i	1+1i	2+1i	0+2i	1+2i	2+2i
0+0i	0+0i	1+0i	2+0i	0+1i	1+1i	2+1i	0+2i	1+2i	2+2i
1+0i	1+0i	2+0i	0+0i	1+1i	2+1i	0+1i	1+2i	2+2i	0+2i
2+0i	2+0i	0+0i	1+0i	2+1i	0+1i	1+1i	2+2i	0+2i	1+2i
0+1i	0+1i	1+1i	2+1i	0+2i	1+2i	2+2i	0+0i	1+0i	2+0i
1+1i	1+1i	2+1i	0+1i	1+2i	2+2i	0+2i	1+0i	2+0i	0+0i
2+1i	2+1i	0+1i	1+1i	2+2i	0+2i	1+2i	2+0i	0+0i	1+2i
0+2i	0+2i	1+2i	2+2i	0+0i	1+0i	2+0i	0+1i	1+1i	2+1i
1+2i	1+2i	2+2i	0+2i	1+0i	2+0i	0+0i	1+1i	2+1i	0+2i
2+2i	2+2i	0+2i	1+2i	2+0i	0+0i	1+2i	2+1i	0+1i	1+1i

·	0 + 0i	1 + 0i	2 + 0i	0 + 1i	1 + 1i	2 + 1i	0 + 2i	1 + 2i	2 + 2i
0 + 0i	0 + 0i	0 + 0i	0 + 0i	0 + 0i	0 + 0i	0 + 0i	0 + 0i	0 + 0i	0 + 0i
1 + 0i	0 + 0i	1 + 0i	2 + 0i	0 + 1i	1 + 1i	2 + 1i	0 + 2i	1 + 2i	2 + 2i
2 + 0i	0 + 0i	2 + 0i	1 + 0i	0 + 2i	2 + 2i	1 + 2i	0 + 1i	2 + 1i	1 + 1i
0 + 1i	0 + 0i	0 + 1i	0 + 2i	2 + 0i	2 + 1i	2 + 2i	1 + 0i	1 + 1i	1 + 2i
1 + 1i	0 + 0i	1 + 1i	2 + 2i	2 + 1i	0 + 2i	1 + 0i	1 + 2i	2 + 0i	0 + 1i
2 + 1i	0 + 0i	2 + 1i	1 + 2i	2 + 2i	1 + 0i	0 + 1i	1 + 1i	0 + 2i	2 + 2i
0 + 2i	0 + 0i	0 + 2i	0 + 1i	1 + 0i	1 + 2i	1 + 1i	2 + 0i	2 + 2i	2 + 1i
1 + 2i	0 + 0i	1 + 2i	2 + 1i	1 + 1i	2 + 0i	0 + 2i	2 + 2i	0 + 1i	2 + 0i
2 + 2i	0 + 0i	2 + 2i	1 + 1i	1 + 2i	0 + 1i	2 + 0i	2 + 0i	1 + 0i	3 + 2i

b) Inverse pairs are as follows:

x	x^{-1}
1 + 0i	1 + 0i
2 + 0i	2 + 0i
0 + 1i	0 + 2i
1 + 1i	2 + 1i
2 + 1i	1 + 1i
0 + 2i	0 + 1i
1 + 2i	2 + 2i
2 + 2i	1 + 2i

c) Let $a = 1 + 1i$. Calculating a^k for $k = 0, 1, 2, \ldots, 9$ gives this:

k	a^k
0	1 + 0i
1	1 + 1i
2	0 + 2i
3	1 + 2i
4	2 + 0i
5	2 + 2i
6	0 + 1i
7	2 + 1i
8	1 + 0i
9	1 + 1i

Notice that a^8 returns to $1 + 0i$ and after that the pattern repeats. Hence rows 0 through 7 of this chart give an isomorphism between $(\mathbb{F}_9^*, \cdot)$ and $(\mathbb{Z}_8, +)$.

d) Appending i to $\mathbb{Z}_5$ does not give a field because $(2 + 1i) \cdot (2 + 4i) = 0 + 0i$ implying that neither $2 + 1i$ nor $2 + 4i$ has a multiplicative inverse.

Chapter 10

10.1 Here's why $(x \oplus y) \odot (x \oplus y) = (x \odot x) \oplus (y \odot y)$.

Consider first the case $x \leq y$.

- On the one hand, $x \oplus y = \min\{x, y\} = x$. Therefore $(x \oplus y) \odot (x \oplus y) = x \odot x = x + x = 2x$.
- On the other hand, $(x \odot x) \oplus (y \odot y) = (2x) \oplus (2y) = 2x$ because $2x \leq 2y$.

The argument when $x \geq y$ is analogous; both $(x \oplus y) \odot (x \oplus y)$ and $(x \odot x) \oplus (y \odot y)$ equal $2y$.

10.2 The equation $x = x + 1$ is not solvable in $\mathbb{R}$, but has two solutions in $\overline{\mathbb{R}}$: ∞ and $-\infty$.

10.3 We calculate the product by expanding all the cross terms:

$$\begin{aligned}
(a + bi + cj + dk)(a - bi - cj - dk) &= a(a - bi - cj - dk) \\
&\quad + bi(a - bi - cj - dk) \\
&\quad + cj(a - bi - cj - dk) \\
&\quad + dk(a - bi - cj - dk) \\
&= (a^2 - abi - acj - adk) \\
&\quad + (bai - b^2i^2 - bcij - bdik) \\
&\quad + (caj - cbji - c^2j^2 - cdjk) \\
&\quad + (dak - dbki - dckj - d^2k^2) \\
&= (a^2 - abi - acj - adk) \\
&\quad + (abi + b^2 - bck + bdj) \\
&\quad + (acj + bck + c^2 - cdi) \\
&\quad + (adk - bdj + cdi + d^2) \\
&= a^2 + b^2 + c^2 + d^2.
\end{aligned}$$

This implies $(1 + i + j + k)(1 - i - j - k) = 4$ and so we have $(1 + i + j + k)^{-1} = \frac{1}{4} - \frac{1}{4}i - \frac{1}{4}j - \frac{1}{4}k$.

10.4 Simple matrix calculations show that

$$M_i^2 = M_j^2 = M_k^2 = M_i M_j M_k = -I$$

where I is the 4×4 identity matrix. This corresponds exactly to the (minimal) defining relation $i^2 = j^2 = k^2 = ijk = -1$.
The value of $\det M_q$ is $(a^2 + b^2 + c^2 + d^2)^2$.

10.5 As in Exercise 10.4, it is easy to check that $M_i^2 = M_j^2 = M_k^2 = M_i M_j M_k = -I$ where, in this case, I is the 2×2 identity matrix.
$\det(M_q) = a^2 + b^2 + c^2 + d^2$ where $M_q = aI + bM_i + cM_j + dM_k$.

10.6 a) Subtract 1 from $\ldots 444444_{\text{FIVE}}$ and we get $-2 = \ldots 4444443_{\text{FIVE}}$.

b) Divide $\ldots 444444_{\text{FIVE}}$ by 2 and we have $-\frac{1}{2} = \ldots 222222_{\text{FIVE}}$. Then add 1 to both sides to get $\frac{1}{2} = \ldots 222223_{\text{FIVE}}$.

10.7 a) 10111_{TWO}.

b) $\ldots 1111111111111111111111111101001_{\text{TWO}}$.

c) 1111111111111111111111111101001.

The point, of course, is that these last two answers are (nearly) the same.

10.8 $a^2 = \ldots 61526043560000000002_{\text{SEVEN}}$. This is very close to 2. The number a is a square root of two in $\mathbb{Q}_7$.

10.9 In base five, the thirteen rightmost digits of $(223032431212_{\text{FIVE}})^2$ are all 4s, so $a^2 = \ldots 4444444444444_{\text{FIVE}} = -1$. That is, $a = \sqrt{-1}$ in $\mathbb{Q}_5$.

10.10 $\ldots 999999_{\text{TEN}} = -1$. There are two ways to see this.

Let $x = \ldots 999999_{\text{TEN}}$. Then we calculate $x - 10x$ as follows:

$$\begin{array}{rl} x = & \ldots 9999999_{\text{TEN}} \\ - \quad 10x = & \ldots 9999990_{\text{TEN}} \\ \hline -9x = & \ldots 0000009_{\text{TEN}} \end{array}$$

and therefore $x = -1$.

Alternatively, add 1 to $\ldots 99999999_{\text{TEN}}$ and notice that the result is $\ldots 00000000_{\text{TEN}}$.

10.11 Note that

$$N = 9879186432 \cdot 8212890625 = 81136677630000000000,$$

which has ten zeros at the end. Therefore $|N|_{10}$ is extremely close to 0. Although neither a nor b is zero, their product $a \cdot b = 0$ in $\mathbb{Q}_{10}$. Numbers such as a and b are called *zero divisors*.

If p is a prime, then the p-adic numbers, $\mathbb{Q}_p$, form a field and there are no zero divisors. See Proposition 0.4 and Exercise 0.8.

Bibliography

[1] R. L. Goodstein. The definition of number. *Mathematical Gazette*, 41(337):180–186, 1957.

[2] Fernando Q. Gouvêa. *p-adic Numbers: An Introduction,* 3rd ed. Springer, 2020.

[3] The Institute of Electrical and Electronics Engineers. *American National Standard Mathematical Signs and Symbols for Use in Physical Sciences and Technology*, 1993.

[4] International Organization for Standardization. *Quantities and Units Part 2: Mathematical Signs and Symbols to be Used in the Natural Sciences and Technology*, 2019. ISO 80000–2:2019.

[5] Donald Knuth. *Surreal Numbers.* Addison-Wesley, 1974.

[6] T. W. Körner. *Where Do Numbers Come From?* Cambridge University Press, 2020.

[7] Carl E. Linderholm. *Mathematics Made Difficult.* Wolfe Publishing, 1971.

[8] Walter Rudin. *Principles of Mathematical Analysis,* 3rd ed. McGraw-Hill, 1976.

[9] Edward Scheinerman. *Mathematical Notation: A Guide for Engineers and Scientists*. CreateSpace, 2011.

[10] Edward Scheinerman. *Mathematics: A Discrete Introduction,* 3rd ed. Cengage, 2012.

[11] Michael Spivak. *Calculus.* W. A. Benjamin, Inc., 1967.

[12] Zach Weinersmith. *Saturday Morning Breakfast Cereal*, A new method. https://www.smbc-comics.com/comic/a-new-method.

[13] Ittay Weiss. Survey article: The real numbers – a survey of constructions. *Rocky Mountain Journal of Mathematics*, 45(3):737–762, 2015.

[14] Eric Weisstein. *Wolfram MathWorld*. https://mathworld.wolfram.com/.

[15] *Wikipedia*: History of quaternions. https://en.wikipedia.org/wiki/History_of_quaternions.

[16] *Wikipedia*: Turtles all the way down. https://en.wikipedia.org/wiki/Turtles_all_the_way_down.

Index

$\mapsto$, 20
$\aleph_0$, 157
$\cap$, 9
$[\![\ ]\!]$, 13
$\cup$, 9
$\div$, 42
$\equiv$, 27
$\in$, 8
∞, 153
$\notin$, 8
$\odot$, 155
$\oplus$, 155
$\otimes$, 58
$\prec$, 149
$\subseteq$, 9
$\triangle$, 17
$\times$, 18
$\varnothing$, 8
$|$, 42

algebraic closure, 148
antisymmetric property, 38
associative property, 34
Augustine of Hippo, xiv
automorphism, 131
axioms, Peano, 40

base notation, 50, 131, 161
basic object, 25, 30, 180
bisection method, 115, 128
boldface, 28
Bond, James Bond, 199
bounded, 120

$\mathbb{C}$, 135
$\overline{\mathbb{C}}$, 154
C_r, 144
cancellation property, 34, 65
Cantor, Georg, 157
cardinal numbers, 157
Cartesian product, 18
Cauchy, Augstin-Louis, 106
Cauchy sequence, 106, 164
closed, 84
closure, algebraic, 148
closure property, 34, 37
commutative
    property, 34, 37
    ring, 61, 65, 70, 75
complement, 9
complete, 133
complete ordered field, xv, 99, 100, 119, 129
completeness property, 98, 118
composite, 43
concatenation, 17
continuum, 97
converge, 133
correspondence, one-to-one, 12

Dedekind, Richard, 83
dense, 79
difference
    set, 9
    symmetric, 17
disjoint, 9
distributive property, 37

209

divides, 42
Division Algorithm, 43
divisor, 42
    zero, 71, 205
doubleton, 17

e, 150
element, 8
empty set, 8
equinumerous, 26
equivalence
    class, 13
    relation, 13
Euclid, 43, 45
Euler's number, 150
eventually, 108
existence, 12
extended real numbers, 153

factor, 42
Fibonacci sequence, 5, 171
field, 76
    complete ordered, xv, 99, 100, 119, 129
    ordered, 78, 149
    skew, 160
Fifteen game, 123
finite, 28
Frege, Gottlob, 24
function, 20
    rational, 80
Fundamental Theorem
    of Algebra, xvi, 147
    of Arithmetic, 43

$\mathbb{H}$, 160
Hamilton, William Rowan, 159
Hanks, Tom, 14
harmonic number, 121
hyperreal numbers, 156

identity element, 21, 34, 37
induction, 41
infinity, 153
    complex, 154

intersection, 9
inverse relation, 18
isomorphic, 123

Julia language, 155

Kronecker, Leopold, 23, 129
Kuratowski, Kazimierz, 28

least upper bound, 97, 118
left ray, 83
    closed, 84
    open, 84
lexicographic order, 149
list, 10
logician, 24
lub, 97, 118

magic square, 124
magnitude, 140
mathematical induction, 40, 41

$\mathbb{N}$, 7
natural number, 7, 23, 30
negative, 77
Newton's method, 122
next, 40
notation
    place-value (base), 50, 131, 161
    scientific, 132
null set, 8
number
    Fibonacci, 5, 171
    harmonic, 121
    natural, 7, 23, 30
    prime, 42
    rational, 74

object, basic, 25, 30, 180
one-to-one correspondence, 12
open, 84
operation, 21
ordered field, 78, 149
ordered pair, 10
ordinal numbers, 158

Index   211

p-adic numbers, 160
pair, ordered, 10
pairwise disjoint, 15
partition, 15
parts, 15
Peano axioms, 40
Peano, Giuseppe, 40
phase, 140
pickle, 180
place-value notation, 50, 131, 161
positive, 77
power set, 157
prime, 43
prime number, 42
product
    Cartesian, 18
proper subset, 10
property
    antisymmetric, 38
    associative, 34
    cancellation, 34, 65
    closure, 34, 37
    commutative, 34, 37
    completeness, 98, 118
    distributive, 37
    reflexive, 13, 38
    symmetric, 13
    transitive, 13, 38
    trichotomy, 48
    well-ordering, 40, 41, 49, 158
Pythagorean Theorem, 82
Python, 168

$\mathbb{Q}$, 74
$\mathbb{Q}_p$, 166
quadratic formula, 143
quadratic residue, 72, 185
quaternion, 159
quotient, 42

$\mathbb{R}$, 85, 111
$\overline{\mathbb{R}}$, 153
$^*\mathbb{R}$, 156
$\mathbb{R}[x]$, 137

radix point, 161
rational
    function, 80
    number, 74
ray, left, 83
recursive, 42
recursive definition, 28
reflexive property, 13, 38
relation, 10
    equivalence, 13
    formal definition, 19
    inverse, 18
remainder, 42
representative, 14, 31
    disjoint, 32
ring, commutative, 61, 65, 70, 75
Robinson, Abraham, 156
Russell, Bertrand, 180

scientific notation, 132
sequence, 105
    Cauchy, 106, 164
    Fibonacci, 5, 171
set, 8
    builder notation, 14, 18
    complement, 9
    difference, 9
    doubleton, 17
    empty, 8
    finite, 28
    null, 8
    power, 157
    singleton, 8
    symmetric difference, 17
singleton, 8
skew field, 160
subset, 9
    proper, 10
subtraction, 62
successor, 40
symmetric
    difference, 17
    property, 13

tic-tac-toe, 124
transfinite
    cardinals, 157
    ordinal, 158
transitive property, 13, 38
triangle inequality, 110
trichotomy property, 48, 60
tropical arithmetic, 154
two's complement, 168

union, 9
uniqueness, 12
upper bound, 97, 118
    least, 97, 118

well-ordering property, 40, 41, 49,
        158

$\mathbb{Z}$, 53, 61
$\mathbb{Z}^*$, 73
$\mathbb{Z}_m$, 67
$\mathbb{Z}_m^*$, 125
$\mathbb{Z}[x]$, 65, 80
$\mathbb{Z}(x)$, 80
zero divisor, 71, 205

Printed in the United States
by Baker & Taylor Publisher Services